Sensory Neuroscience: Four Laws of Psychophysics

Jozef J. Zwislocki

Sensory Neuroscience: Four Laws of Psychophysics

 Springer

Dr. Jozef J. Zwislocki
Syracuse University
L.C. Smith College of Engineering
and Computer Science
621 Skytop Rd.
Syracuse NY 13244
USA
joe_zwislocki@isr.syr.edu

ISBN: 978-0-387-84848-8   e-ISBN: 978-0-387-84849-5
DOI: 10.1007/978-0-387-84849-5

Library of Congress Control Number: 2008940434

springer.com

Preface

Hopefully, this book may be of interest not only to psychophysicists who formally call themselves "psychophysicists," and of whom there are very few but also to all those who use psychophysics in their activities, often not even realizing they do. Every educated person has at least an idea of what physics is, the same goes for psychology. But what is psychophysics? The name suggests a combination of both. More specifically, psychophysics is a science concerning human and animal sensory responses to physical and also chemical stimuli. Seeing is a response to light. We may see the light as red or yellow or any other color or a mixture of colors. It may appear as a short flash or a steady illumination; it may be patterned and appear as stripes, circles, squares, or any other shapes. Similarly, hearing is a response to sound that may appear as noise, musical sound, or a pure tone to pick a few examples. Tactile feeling results from pressure on the skin surface, feeling of warmth from heat, smelling from odorant substances, taste from gustatory chemicals, and so on.

If psychophysics deals so generally with sensory responses, why are there so few psychophysicists? Simply, because they tend to specialize in particular senses, like vision or hearing or taste, for example, and call themselves visual scientists, auditory scientists, and so on.

Psychophysics has wide-ranging applications in our professions – in medical diagnostics, most specifically, in ophthalmology and otology, in optometry and audiology, also in engineering, in architecture, in the arts, even in such fields as traffic control (think of traffic lights, for example).

Psychophysical processes are involved in almost everything we do. When we compare the height of a house to another by eye – that is psychophysics; when we compare the pitch of one sound to another – that is psychophysics also; even when we decide which coffee or tea we prefer, we perform a psychophysical tasting operation. In sports, whenever we throw or catch or hit a ball – subjective psychophysical measurement is involved. We can hardly make a move without running into psychophysics.

Psychophysics was discovered, or invented, if you prefer, by a nineteenth century physicist, Gustav Theodor Fechner, who was interested in the relationship between the material and the spiritual. In modern times, it has become a science of our, or even animal, sensory responses to physical or chemical stimuli, as I already stated. To a large extent, psychophysics is a quantitative science. For example, it attempts to describe numerically how fast a sensation grows with the intensity of a physical stimulus. When we increase the intensity of sound, how fast does its loudness increase, by how much do we have to increase the intensity of a light to increase its brightness by a factor of two or three or any other ratio? It also attempts to specify thresholds of detectability. What is the smallest pressure on the skin we can just notice, or the smallest perceptible concentration of an odorant or a gustatory chemical?

Psychophysics, as established by Fechner, aimed at finding pervasive quantitative rules for such relationships. In physics, which derives a tremendous power from them, such rules are called scientific laws. The first psychophysical law was established by Fechner on the basis of the observation of Ernst H. Weber that a just noticeable increment in lifted weights is directly proportional to the base weight. The observation proved to hold not only for weight but also for other physical variables as well. Fechner established yet another famous law named after him. On an intuitive insight, he decided that subjective impressions grew as logarithms of the physical stimuli that produced them. The law proved much later to be incorrect. Nevertheless, together with Weber's law, it demonstrated the importance Fechner, as a physicist, attached to scientific laws. They point to relationships of great generality from which specific relationships can be derived. In so doing, they provide a basic structure of a science.

Psychophysics has been founded in mid-nineteenth century by Gustav Theodor Fechner, a German scientist and philosopher with a strong mystical bend, who begun his scientific career as a physiologist but, subsequently, became a self-taught physicist with a consuming interest in the relationship between body and mind. Psychophysics resulted from this interest and, in practice, concerned mainly sensations, such as those of brightness and loudness or heaviness. Undoubtedly, under the influence of Fechner's physics background, it was conceived as a quantitative science structured around general laws. Two became famous, Weber's law conceived by Fechner on the basis of Weber's experiments with lifted weights, according to which a just noticeable weight difference was directly proportional to the reference weight, and Fechner's law, not based on an experiment but on an intuitive insight of Fechner's, according to which subjective magnitudes of sensations grew as logarithms of the intensity of stimuli that evoked them. The logarithmic function seemed appropriate because it reflected the subjectively slow growth of such sensations as brightness by comparison to the light intensity that produced it. Fechner, like his contemporaries, did not believe that sensations can be quantified experimentally, and I have never ceased wandering how, if this were true, he and some others before him were able to feel that sensations grew less rapidly than the corresponding physical intensities that, of course, were measured instrumentally. As discussed in this monograph, Weber's law, at least in a somewhat modified form, survived the test of time. This is not true for Fechner's logarithmic law that became replaced by a power law based on ample experimental documentation.

Foundation of psychophysics preceded by about a decade the establishment of the laboratory of experimental psychology by Wilhelm Maximilian Wundt and partially inspired it. The laboratory gave rise to several branches of psychology. Psychophysics evolved to become one of them; a branch that has been both pervasive with respect to the others and, at the same time, distinct because of its strong interactions with physics and chemistry, and because of its many applications outside psychology. Some of the most famous psychophysicists begun their research careers as physicists or engineers. Among the best known applications have been those to medical diagnostics, most specifically, in ophthalmology and otology, in optometry and audiology, others to engineering, architecture, the arts, even to such fields as traffic control – think, for example, of the colors of traffic lights.

Psychophysical research is often performed in tandem with physiological research, in particular, with neurophysiological one. Psychophysics attempts to determine what our and animal sensory systems do, physiology, how they do it.

Most recently, psychophysics and physiology together have become essential parts of environmental sciences by telling us how our environments affect us.

Psychophysical processes are involved in almost everything we do. When we compare the height of a house to another by eye – that is psychophysics; when we compare the pitch of one sound to another – that is psychophysics also; even when we decide which coffee or tea we prefer, we perform a psychophysical tasting operation. In sports, whenever we throw or catch or hit a ball – subjective psychophysical measurement is involved. We can hardly make a move without running into psychophysics.

I have watched psychophysics evolve during four to five decades. At and participated actively in its evolution, at first, as a young electrical engineer who chose to work in otology on diagnostic procedures rather than pursue a traditional engineering career. The place was the Department of Oto-Rhino-Laryngology at the Medical School of the University of Basel, Switzerland. During that time, I discovered what is now called "forward masking," a phenomenon according to which a preceding tone decreases the audibility of a following tone for a short span of time, and that only recently experienced a wave of research popularity. I discovered that, in the absence of direct acoustic interference, a sound in one ear masked in a strongly time-dependent and frequency-dependent fashion the audibility of sound in the contralateral ear. A phenomenon I called "central masking" because it had to take place in central neural interaction. I also discovered that the size of just noticeable differences in tone intensity depended on loudness rather than directly on the intensity – a revolutionary, paradoxical appearing finding, inconsistent with the foundations of stimulus-oriented psychophysics. It led to a diagnostic method concerning inner ear disorders.

A doctoral dissertation written at that time provided me with the degree of a Science Doctor and brought me to Harvard's Psychoacoustic Laboratory where I spent 6 years. The laboratory belonged to the Department of Experimental Psychology, and I suddenly found myself, surrounded by psychologists. It is in this ambiance that, together with several coworkers, I demonstrated an unexpectedly strong dependence of the measured threshold of audibility for pure tones on practice and motivation of the listeners. On this occasion, taking off from the automated audiometer of Georg v. Békésy's (Nobel, 1960) I devised a "criterion-free" automated method of threshold determination. The method, based on somewhat different statistics, was later reinvented twice, once in hearing and once in vision. All these methods now form the class of adaptive methods for automatic, criterion-free threshold determination.

My work on practice and motivation and the criterion-free method fitted right into the "theory of signal detectability," which burst upon psychophysics around that time and provided a mathematical foundation for criterion-free signal detection. Mainly through the efforts of David M. Green, it revolutionized the part of psychophysics, especially auditory psychophysics, which dealt with thresholds of detectability.

The theory of signal detectability did not have much effect on measurement of suprathreshold events. Such an effect was provided by a discovery made by Stanley Smith Stevens, the director of the Psychoacoustic Laboratory. He found that people without any special training were able to express ratios between the subjective magnitudes of their sensations in numbers. For example, they were able to tell how many times a given sound was louder than another sound, or, a given light brighter than another light. From their numerical responses, Stevens was able to construct functions relating the sensation magnitudes to stimulus magnitudes. He found that, almost invariably, the functions conformed to power functions. Through extensive experimentation, he determined that the generality of this phenomenon was so extensive that it deserved to be regarded as a scientific law. The law is now generally known in psychophysics as Stevens' Power Law.

Soon after the establishment of Stevens' law, my research tenure at Harvard was over, and I accepted a Faculty position at Syracuse University. Here, following up on my experience in diagnostic otolaryngology and at Harvard's Psychoacoustic Laboratory, I organized successively the Bioacoustic Laboratory and the Laboratory of Sensory Communication that advanced to become the Institute for Sensory Research at the departmental level. The Institute, as well as its precursor pioneered multidisciplinary research on human and animal senses. Its multidisciplinary faculty included the disciplines of anatomy, neurophysiology, and psychophysics in three sense modalities: hearing, touch, and vision. The multisensory nature of the Institute promoted intersensory comparisons, which led to the discovery of some fundamental intersensory generalities.

This book has resulted from the experience I acquired during my academic career at in all three places, Basel University, and Harvard University, but mainly, at the Institute for Sensory Research at Syracuse University. It also resulted from the conviction that scientific laws form the backbone of a science. Thanks in part to the multisensory nature of the Institute, the laws I have been able to propose apply, with some exceptions, to most if not all human senses. No laws applying exclusively to one or another sense modality have been included.

At the end of this preface, I want to abandon psychophysics for a more personal subject and express my most sincere thanks to two persons who contributed in two different but essential ways to this book. As custom dictates, I first thank

Nicole Sanpetrino, my graduate assistant, for her dedicated and excellent help in the graphics, the editing, and the indexing of this book. Without her help, my task of putting this book together would have been infinitely more difficult. Last but not the least, I thank my wife, Marie, for her inspiration, constructive criticism, and patiently putting up with me while I was being engulfed by this book. Before all, I thank her for keeping me all together during the arduous task of writing.

Syracuse, New York   *Josef J. Zwislocki*

Contents

Introduction

Scientific laws are defined here as quantitative, invariant relationships of broad generality. Newton's law $F = am$, where F means a force applied to the mass, m, to produce an acceleration, a, and Einstein's $E = mc^2$, where E means energy, m, mass, and c, the speed of light are classical examples. The laws constitute the backbone of a science from which other relationships can be derived, and the effects of variables producing deviations from the relationships defined by the laws studied. For example, Newton's law applies to objects falling in vacuum. When an object falls in air due to the effect of gravity, its acceleration is decreased by air resistance. To establish a law empirically, all the variables not specified in the law must be eliminated or their effects determined and accounted for.

Here, the concept of scientific laws is applied to psychophysics, a science established by Gustav Theodor Fechner, a nineteenth-century German physicist and physiologist, to deal with the relationship between the spiritual world and the material world. More down to earth, psychophysics may be defined nowadays as a science of quantitative relationships between psychological variables and the physical variables that elicit them. "Physical" is used here as a generic term including "chemical." Some of the relationships are so intimate that before sufficient instrumentation was developed, the only way people knew about the physical events was through their senses. This is probably the reason why, even today, the same word is used for the physical light as for the sensation of it. The same is true for sound and some other physical variables. We have to be clear in specifying whether we mean the physical variable or its sensation. When we say: "the light is bright" we really mean our sensory impression, not the physical quantity that we can know only through inference. The inference may be quite inaccurate and depend on context variables. Optical illusions are well known.

Before the relationships between the psychological variables and their underlying physical variables could be determined, methods and instrumentation for independent measurement of both had to be established. This has been accomplished for the physical variables under a satisfactory number of circumstances. What about the measurement of psychological variables? It remains controversial, and the dispute, begun for good in the nineteenth century, goes on. The fundamental question continues to be: can subjective variables, such as brightness or loudness or pressure or sweetness or all the other sensations we can imagine, be measured? Most people seem to think in a mystical way that they are part of our inner world that is not accessible to others, therefore, not measurable. For this reason, most of the psychophysics pursued today avoids specifying the magnitudes of these variables and limits itself to addressing human observers as null instruments. Experiments are limited to questions of detectability – did an event occur or not? In a pair of unequal events, was the greater or smaller event presented? In classical psychophysics that preserves some elements of subjectivity, the questions become – did you perceive the event; which event appeared the greater? Still, specification of the absolute magnitudes of the events is avoided.

In psychophysics that restricts the observer, or subject, to a null device; the determination of sensory characteristics occurs indirectly by measuring one stimulus variable as a function of another. For example, the threshold of audibility is determined as a function of sound frequency, the threshold of visibility as a function of the wavelength of light, the vibration detectability as a function of vibration duration, and so forth. It is also possible to measure magnitudes of different stimuli producing equal subjective magnitudes; for example, sound intensities at two different sound frequencies that produce equal loudnesses. All such measurements have proven to be useful. Nevertheless, they are limited to threshold values, or to subjective magnitudes relative to other subjective magnitudes specified indirectly in terms of stimulus values that produce them. Such stimulus-oriented psychophysics provides only an incomplete image of our sensory functioning that most often occurs at suprathreshold stimulus values not referred to specific reference standards.

The situation begun to change cautiously in mid-twentieth century when S.S. Stevens demonstrated more convincingly than his predecessors that mutually consistent ratio measurements of loudness and brightness were possible. The initial experiments were performed in hearing. An observer was given a reference standard consisting of a tone at a predetermined intensity and a number to express its subjective loudness magnitude. He was instructed to assign numbers to subsequently presented tones in proportion to their loudness magnitudes relative to the standard. The numbers proved to follow a power function. Repetition of the experiment on several other observers produced similar responses. An analogous result was obtained when light flashes were substituted for the tonal stimuli. The subjective brightness magnitudes, as expressed by assigned numbers, followed a similar power function. Stevens decided that he may have found a general principle for the relationships between sensory stimulus intensities and the subjective sensation magnitudes they evoked. Because the relationships followed power functions, he designated the principle as the Power Law. The Power Law has been confirmed by many experimenters in many experiments performed in many sense modalities. Next to the Weber Law that withstood the test of almost two centuries, it is the best documented general relationship of psychophysics. Because it may be considered as the answer to Fechner's fundamental question of the relationship between the "spiritual" and the "material," to use Fechner's language, Stevens regarded the Power Law as *the* Psychophysical Law. In this monograph, the view is accepted that the Power Law is the most fundamental law of psychophysics, and I designate it as the First Law of Psychophysics. Nevertheless, additional laws are possible and may have considerable usefulness.

Before the Power Law was firmly established, methods of measuring psychological quantities had to be developed. The original method introduced by Stevens to measure loudness and brightness, which he called "magnitude estimation," proved partially misguided and produced the right results somewhat by lucky coincidence. Stevens and his coworkers soon discovered that the exact functions relating the psychological magnitudes to the underlying stimulus magnitudes depended on the designated reference standards, and the "best" power functions were obtained when the observers were allowed to choose the standards themselves. This discovery suggested that the observers did not obey strictly the rules of ratio scaling, which allow the reference standards to be entirely arbitrary, but, to some degree, attached to numbers absolute values.

A more systematic investigation of the effects of reference standards was performed by Rhona P. Hellman, a graduate student of mine, and myself. The investigation led us to the conclusion that experimental observers do not use numbers in a relative way, depending on chosen units, but rather in an absolute way. In other words, probably because of the way they are used in everyday life and the way children use them when learning them, they acquire absolute subjective values. Children learn numbers by counting objects – pebbles or pencils, for example. As a consequence, coupling of numbers to perceived objects occurs early in life according to the rules of numerosity where numbers have absolute values. These values appear to be extrapolated to continua. When asked to assign numbers to subjective impressions of line length or to loudness, adults and children produce the same absolute functions within the range of the numerals they know. If numbers acquire absolute subjective values, magnitude estimation (ME) becomes a matching operation. The subjective values of numbers are matched to the subjective values of whatever variable is being scaled.

Because Stevens' method of ME appeared to produce biased results, being asymmetrical, he introduced a complementary method in which numbers were given by the experimenter, and the observers had to find matching sensation magnitudes that they produced by manipulating appropriate instrumental controls. He called this method "magnitude production" (MP). In the methods of scaling subjective magnitudes developed by myself, Hellman, and several other coworkers, the numbers are assumed to have absolute subjective values. Consequently, we call what started as ME, "absolute magnitude estimation" (AME), and what started as MP, "absolute magnitude production" (AMP). The designations conserve Stevens' tradition but are not completely accurate because both are regarded as matching operations. The methods have opened a wide world of subjective magnitudes to measurement in spite of objections by staunch conservatives that they do not constitute legitimate measurements. The mutual consistency of the results they produce belies the objections.

Sensation magnitudes almost generally follow power functions of adequate stimulus variables, except at very low values of these variables. As thresholds of detectability of the variables are approached, their subjective magnitudes converge on direct proportionality to the stimulus intensity or a related variable. Line length squared would be an example of such a variable. For sufficient lengths, the subjective line length tends to be directly proportional to the physical line length. As surprising as it may appear, this is no longer true for very short thin lines that become somewhat difficult to see. According to measurements of N.M. Sanpetrino, my Graduate Assistant, the subjective line length then becomes proportional to the square of the physical line length. The phenomenon can be explained by the physiological noise that is added to the visual line image. In agreement with the theory of signal detectability, such a process can be expected to take place near the threshold of detectability of all sensory stimuli. It would be consistent with a linear relationship of subjective magnitudes to stimulus intensity. Because I was able to ascertain empirically such a relationship for many sensory modalities and because of the likely generality of the underlying physiological process, I am suggesting the relationship as the Second Law of Psychophysics.

Additivity of subjective magnitudes is introduced in this monograph as the third law. Its demonstration is essential for the understanding of the function of a sensory system. It signals linear processing. Additivity is also essential in validating the scales of measurement obtained by more direct means, such as AME. Such scales can be constructed by simply adding quantities defined as units. Linear summation of two units produces a quantity equal to two units. Linear summation of two doubled quantities produces a quadrupled quantity, and so forth. Early attempts at establishing the functional relationship between loudness and sound intensity were based on such an additive process on the assumption of additivity. Much of the chapter concerning the additivity is dedicated to the demonstration that it does take place in several, perhaps all sense modalities under appropriate conditions. The demonstration is complicated by the fact that, according to physiological evidence, the summation process is preceded by more peripheral neural processes that may introduce nonlinear interactions. Nevertheless, existence of additivity has been demonstrated with scientific certainty in hearing, touch and vision. The situation in chemical senses had to be left unresolved.

The fourth and last law included in this monograph concerns detectability of intensity increments. In its classical nineteenth-century form of Weber's law, according to which the just detectable intensity increments or, more generally, magnitude increments are directly proportional to the base intensity, or magnitude, it is probably the oldest law of psychophysics. The law is often expressed as the Weber fraction consisting of the ratio between the just noticeable increment and the base magnitude. The fraction tends to have a constant value, except at very low stimulus values, where it rapidly increases. In hearing, for pure tones, and in vibrotaction, for any stimuli, the value tends to decrease slowly as the base intensity is increased. The phenomenon is referred to as the "near miss to Weber's law."

In more recent times, paradoxical-like properties of Weber's law have been discovered in hearing. When measured by means of just detectable intensity increments or the difference between two intensity increments, Weber's fraction has been shown not to depend on the rate of growth of loudness with stimulus intensity but only on the loudness itself. When measured as the standard deviation of the variability of loudness matches between two tones with loudness magnitudes increasing according to two different functions, it has been found to depend on the slopes of the functions in a predictable but complicated way. Counterintuitively, it depended not only on the slope of the loudness function of the ear in which the sound intensity was varied but also on the slope of the loudness function of the contralateral ear. Somewhat unexpectedly, I found it possible to describe the differential intensity sensitivity in all its methodological variations by one simple mathematical equation. I suggest the equation as an expression of a General Law of Differential Intensity Sensitivity.

At the end of this introduction, a crucial caveat must be added. All the mathematical theories used in this monograph are based on the assumption of linear interactions. Nonlinear interactions are excluded. Nevertheless, most sensation magnitudes are compressed functions of underlying stimulus intensities, the powers of the Power Law functions have exponents different from unity, usually, smaller than one. Every physiologist must know that the process of generating neural action potentials is nonlinear, and the rate at which the potentials, usually called "spikes," occur is a saturating function of stimulus intensity. Yet, several researchers have been able to show for hearing that the compression originates in the cochlear mechanics, and I pointed out that, for the most part, it is likely due to a kind of automatic gain control (AGC) that produces negligible distortions of waveforms of sound. AGC is used generally in radio communication. If it produced nonnegligible wave distortions, telecommunication would become impossible. The same goes for the auditory system.

Information transmission through neural spikes can be linear if it occurs through modulation of the spike rate. The overall output of the auditory nerve for pure-tone stimuli has been demonstrated to parallel the loudness function. This suggests overall linear processing above the level of the auditory nerve. The processing does not have to be linear in detail and, according to physiological evidence, it certainly is not. But the nonlinearities have to cancel each other in the summated neural output to produce what is called a "quasi-linear" process.

In the sense of touch, the subjective sensation of pressure is nearly directly proportional to the depression of the skin surface, so that here the problem of compression does not arise. Deviations from linearity in the sensation magnitudes observed in the chemical senses are less substantial than in hearing, but their physiological mechanisms are not clearly specified. In vision, the substantial compression evident in the brightness functions of luminous targets seen on a black background has not been analyzed in this monograph in terms of the theory of linear signal processing, except for small signals, where linear approximations are possible.

Importantly, in all instances, the theoretical results have been validated by empirical confirmation.

Chapter 1
Stevens' Power Law

1.1 Definition and Genesis

According to the power law, sensation magnitudes grow as power functions of stimulus intensities that produce them. The law was first proposed by S.S. Stevens for light and sound. It was announced in a 1953-paper presented before the National Academy of Sciences (USA) (cit. Stevens, 1975). Subsequently, Stevens suggested it as a general law describing quantitatively the relationships between human sensations as well as other subjective impressions and the physical stimuli that evoke them (rev. Stevens, 1975). According to the proposed law, the relationships approximate power functions of the form

$$\psi = k\phi^{\theta} \tag{1.1}$$

with ψ symbolizing the sensation magnitude, ϕ, the magnitude of the physical stimulus, θ the power exponent, and k, a dimensional constant.

The genesis of the law has a stormy history. The question of the relationship between the physical world surrounding us and its representation in our minds has haunted scientists for centuries, but did not mature to a quantitative science until Gustav Theodor Fechner, the physicist becoming philosopher, established the science of Psychophysics in 1860 (English translation, 1966). Before he did, he postulated in 1850 (cit. Stevens, 1975) on an intuitive insight that sensation magnitudes were related to the magnitudes of physical stimuli by logarithmic functions. The logarithmic "formula," as Fechner called it, had an important antecedent. Already in 1738, the famous Swiss mathematician, D. Bernoulli, came to the conclusion that the subjective value of money increased as the logarithm of the amount of money (Bernoulli, translation, 1954). He observed that the subjective value increased much more slowly than the objective one, an impression that was in agreement with the strongly compressed logarithmic function.

Fechner's formula received some experimental support. On the request, of another physicist, J.A.F. Plateau, J. Delboeuf (1873; cit. Stevens, 1975) let some

J.J. Zwislocki, *Sensory Neuroscience: Four Laws of Psychophysics,*
DOI: 10.1007/978-0-387-84849-5_1,
© Springer Science+Business Media LLC 2009

painters mix white and black paints in various proportions to obtain equal appearing contrast steps or intervals. The result followed roughly a logarithmic function. Furthermore, an interval scale of stellar brightness created in antiquity, around 150 BC by the astronomer Hipparchus for classification of stars was found to agree roughly with a logarithmic scale when physical photometry became possible (Jastrow, 1887; cit. Stevens, 1975). Fechner found his formula to be consistent with the experiments of E.H. Weber (1834; cit. Marks, 1974) who had determined that difference limens (DLs) or just noticeable differences (JNDs) between sensory stimulus magnitudes were directly proportional to stimulus magnitudes – a quite general relationship that became known as Weber's fraction, or law. Assuming that the JND steps had equal subjective magnitudes, and integrating the relationship, Fechner recovered the logarithmic formula (cit. Stevens, 1975). To be mathematically correct the derivation should take the form:

$$\Delta\psi = a\frac{\Delta\phi}{\phi} \tag{1.2}$$

After integration,

$$\psi = a\log\left(\frac{\phi}{\phi_0}\right) \tag{1.3}$$

where ψ means the subjective (psychological) magnitude, as before, ϕ the physical magnitude with ϕ_0 as its reference value, and a, a dimensional constant.

Buoyed by these and other similar results, Fechner's formula became regarded as a psychophysical law that reigned supreme for almost a century and even invaded neurophysiology (e.g. Matthews, 1931; Hartline and Graham, 1932). Communication engineers devised a logarithmic scale based on a unit called decibel (dB) to match what they thought would be the loudness function. A difference of 10 dB meant an intensity ratio of 10, that of 20 dB, one of 100, that of 30 dB one of 1,000, and so forth. Because sound intensity is proportional to the square of sound pressure, a sound pressure ratio of 10 is equivalent to 20 dB.

Unexpectedly, the logarithmic decibel scale proved to be mortal to Fechner's law. When the decibels of sound intensity were doubled, for example, from 40 to 80 dB, the subjective loudness did not double as it should have, were it proportional to the logarithm of the intensity, but increased much more (Churcher, 1935; cit. Marks, 1974). In addition, when equal numbers of intensity JNDs were added up at two different sound frequencies, the resulting loudness magnitudes did not appear to be equal, as could be easily verified by a direct loudness match (Newman, 1933; cit. Marks, 1974). The loudness grew faster than predicted by the logarithmic function. This conclusion was confirmed by many other experiments discussed extensively by Marks (1974).

The significance of the mentioned subjective impressions goes far beyond demonstrating that Fechner's logarithmic law cannot be true. They suggest that sensations have quantifiable magnitudes. This insight is consistent with Bernouilli's and Fechner's decisions to use highly compressive functions in describing the growth of the subjective magnitudes they experienced, rather than a linear one. Other scientists, who may not have used logarithmic functions for similar purposes, used compressive functions, nevertheless. They all must have been able to gauge subjectively the rate of growth of their sensation magnitudes.

The above observation makes the outcry of some prominent psychologists and philosophers against the attempts of Fechner and a few others to quantify sensation magnitudes appear hollow. To quote from James (1890), W. Wundt, the father of experimental psychology stated: "How much stronger or weaker one sensation is than another, we are never able to say." Of course, Wundt thought of numerical ratios which may not be explicit in the feeling that loudness grows with sound intensity more rapidly than according to a logarithmic function. Nevertheless, the logarithmic function is expressed numerically. Again, quoting after James (1890), the famous philosopher, Carl Stumpf, stated: "One sensation cannot be a multiple of another. If it could, we ought to be able to subtract the one from the other, and to feel the remainder by itself. Every sensation presents itself as an indivisible unit." Stumpf's statement was later parodied by two British physicists, Richardson and Ross (1930), who wrote: "One mountain cannot be twice as high as another. If it could, we ought to be able to subtract the one from the other and to climb up the remainder by itself. Every mountain presents itself as an indivisible lump." They went on to produce a numerical scale of loudness. The method they used may be regarded as a precursor of the method Stevens subsequently worked out in great detail and called "magnitude estimation." Their result suggested that loudness was related to sound intensity by a power function rather than a logarithmic function.

Perhaps Richardson and Ross were the first to come up with an empirical power-function relationship between a physical variable and the subjective impression it evoked. According to Stevens (1975), the idea of the power function relationship may have been first conceived by a young eighteenth century mathematician, Gabriel Cramer, however, whose work was cited by Bernoulli. Cramer, like Bernoulli was interested in the subjective value of money.

The decade following Richardson and Ross' paraphrase witnessed an explosion of experiments on loudness quantification, spurred by developments in electronics and communication engineering. Sound intensity and, with it, loudness became easy to vary. Most experiments followed ratio procedures in which the observers were asked to produce loudnesses that were subjectively twice, three times, four times and so on louder than a given standard loudness. Fractionation, in which the loudness had to be made 1/2, 1/3 or smaller than the standard was also employed. Many of the results were summarized by Churcher (1935) and used by Stevens (1936) to construct a loudness scale – the so-called *sone* scale. One sone served as a unit and was defined as the loudness of a 1,000 Hz. tone presented binaurally 40 dB above its threshold. The resulting function clearly departed form a logarithmic one and resembled a power function instead.

The ratio procedures, which required an intensity adjustment by the observer, were tedious and time consuming. Stevens looked for a more efficient method. He stumbled upon one almost accidentally in 1953, a year that may become almost as important for psychophysics as the year 1850 in which Fechner conceived of quantifying sensation magnitudes (cit. Stevens, 1975). Stevens' discovery had such an impact on psychophysics that I would like to honor it here by quoting verbatim his anecdotal description of it (Stevens, 1975).

> What I wanted was a method that would tell me the overall form of the scale, from a weak stimulus to a strong one. I expressed that idea to a colleague who objected that he had no loudness scale in his head from which he could read such values directly. That was a novel thought, however, and I persuaded him to explore it with me.
>
> I turned on a very loud tone at 120 decibels, which made my colleague jump, and which we agreed would be called 100. I then turned on various other intensities in irregular order, and for each stimulus he called out a number to specify the loudness. I plotted the numbers directly on a piece of graph paper in order to see immediately what course was being followed by the absolute judgments, as I first called them. The experiment seemed to work so well that I proceeded to enlist other observers. The plots of the magnitude estimations, as I now call them, that were produced by the first half-dozen observers are shown in Fig. 7 (Fig. 1.1 here). Each observer's estimations were plotted on a separate graph, I had no assurance that it would be proper to average the data from different observers. The general agreement among the responses of the first few observers persuaded me that I had probably hit upon a promising method, and that the potential of the procedure ought to be explored seriously.

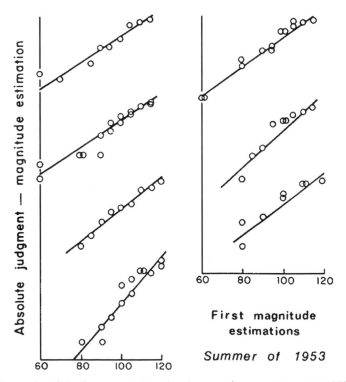

Fig. 1.1 The results of the first magnitude-estimation experiment performed in 1953. The data points indicate single estimates given by individual observers relative to a reference sound pressure level of 120 dB assigned the number 100. Reproduced from Stevens (1975) with permission from John Wiley & Sons, Inc.

The individual data resulting from this first of Stevens' experiments on magnitude estimation, plotted in Fig. 1.1, follow roughly power functions. They are indicated by the straight lines on the double logarithmic coordinates of the graphs. The slopes of the lines, which indicate the exponents of the power functions, are all similar, although moderate individual differences are apparent. Subsequent experiments on magnitude estimation convinced Stevens that the deviations from median values were random and warranted inter-individual averaging. The distributions were approximately normal on log coordinates, so that the geometric mean was accepted as a measure of the central tendency.

In addition to the auditory experiments, Stevens applied the method of magnitude estimation to vision in an attempt at determining the magnitude of brightness. In the first experiment the stimulus consisted of a circular milk- glass surface mounted in a black screen and illuminated from behind. A group of 18 observers participated. The medians of their data followed a power function with a cube root exponent, similar to that for loudness, as shown in Fig. 1.2. Stevens concluded that "Perhaps both brightness and loudness are governed by the same simple power law." He performed some confirmatory measurements and, in the same year, presented a paper before the National Academy of Sciences "suggesting that a cube-root law governs the sensory response to light and sound" (Stevens, 1953).

To test the sensory generality of the power law, Stevens and his coworkers applied the method of magnitude estimation to a large number of sense modalities. Almost always, the result was fundamentally the same – the estimated magnitude grew as

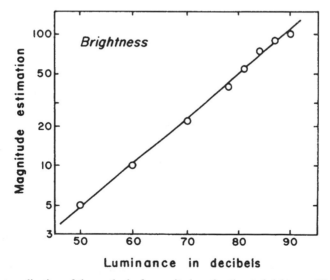

Fig. 1.2 First application of the method of magnitude estimation to brightness. The data points represent medians of 2 estimates by 18 observers, referred to a standard high luminance called 100. The straight line fitted to the data on the log-log coordinates follows a cube-root power function. Reproduced from Stevens (1975) with permission from John Wiley & Sons, Inc.

a power function of the physical magnitude, although the power exponent varied, depending on the sense modality and some stimulus conditions (e.g. Marks 1974; Stevens, 1975).

1.2 Theory of Magnitude Estimation

A quantitative empirical law must rest upon bona fide accurate measurements. Does magnitude estimation afford such measurements? By assigning the number "100" to the loudness produced by a 1,000-Hz tone at 120 dB sound pressure level (SPL), Stevens, in effect, defined arbitrarily a unit of measurement. Theory of measurement allows for an arbitrary definition of units on what is called ratio scales, provided the transformations leave the ratios invariant (e.g. Stevens, 1946, 1951). For example, we can measure the length in centimeters, meters, inches, feet or still other units.

In a more systematic experiment than the initial ones, Stevens chose the number 10 associated with SPLs of either 80 or 90 dB as reference standards and computed population medians that do not depend on the distribution of individual responses (Stevens, 1955, 1956). He instructed 18 observers to assign numbers to loudness magnitudes of 1,000-Hz tone bursts produced at randomly chosen SPLs so as to express the perceived ratios between these magnitudes and the standard magnitude. In the 1956 article entitled: *"The direct estimation of sensory magnitudes – loudness,"* Stevens gives a detailed description of his instructions to observers who sat before a pair of switches, one for producing the reference tone and the other, the variable tone through a pair of earphones:

> The left key presents the standard tone and the right key presents the variable. We are going to call the loudness of the standard 10 and your task is to estimate the loudness of the variable. In other words, the question is: if the standard is called 10 what would you call the variable? Use whatever numbers seem to you appropriate – fractions, decimals, or whole numbers. For example, if the variable sounds seven times as loud as the standard, say 70. If it sounds one fifth as loud, say 2: if a twentieth as loud, say 0.5, etc. Try not to worry about being consistent; try to give the appropriate number to each tone regardless of what you may have called some previous stimulus. Press the 'standard' key for 1 or 2 s. and listen carefully. Then press the 'variable' for 1 or 2 s. and make your judgment. You may repeat this process if you care to before deciding on your estimate.

The experimental results are shown in Fig. 1.3. They are representative of intended ratio scaling, because the observers listened alternately to the fixed reference standard and the variable stimuli and could estimate directly the loudness ratios between them. Had the observers been able to estimate the ratios accurately, the resulting data points would have fallen on parallel lines indicated in the figure. However, the data points associated with the 80-dB standard tend to fall above the corresponding line below the standard. Those associated with the 90-dB standard deviate upward from a parallel line at high SPLs. The deviations suggest that the ratio estimates were not completely independent of the implied units of measurement, and the criterion of ratio scaling was poorly satisfied. The deviations are not large but, attempting to investigate their nature, Stevens was able

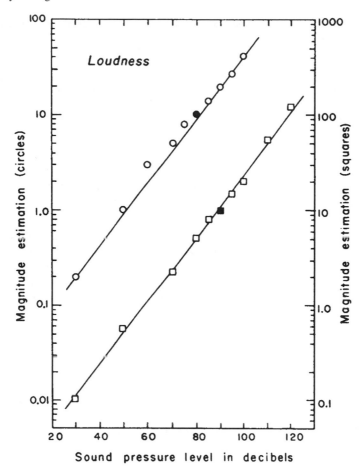

Fig. 1.3 Magnitude estimation of loudness of a 1,000-Hz tone with reference to a standard number 10 assigned to either a SPL of 80 or 90 dB. The data points represent medians of two estimates by 18 observers; the straight lines follow power functions with an exponent of 1/3. Reproduced from Stevens (1956) used with permission of the author and the University of Illinois Press. Copyright 1956 by the Board of Trustees of the University of Illinois

to produce much larger ones by more substantial changes in reference standards (Stevens, 1956). Although his experiments indicated that the observers involved in them were not able to follow accurately the ratio-scaling instructions, Stevens was reluctant to abandon the reference standard entirely. After all, magnitude estimation evolved from ratio-scaling procedures, such as halving and doubling, which always involved a reference standard. Nevertheless, on the suggestion of his wife, as he states it (Stevens 1975), who was one of his most frequent observers, he went so far as to allow his observers to choose their own numbers for the first stimulus presented. They were to choose numbers for subsequent stimuli consisting of 1,000-Hz tones in proportion to the sensation magnitudes they experienced. Here are his instructions (Stevens, 1956):

> I am going to give you a series of tones of different intensities. Your task is to tell me how loud they sound by assigning numbers to them. To turn on the tone you simply press the key. You may press it as often as you like. When you hear the first tone, give its loudness a number – any number you think appropriate. I will then tell you when to turn on the next tone, to which you will also give a number. Try to make the ratios between the numbers you assign to the different tones correspond to the ratios between the loudnesses of the tones. In other words, try to make the numbers proportional to the loudness, as you hear it.

Stevens comments: "All but one of the O's had previously made estimations with the aid of a fixed standard, and these instructions came as a mild shock to some of them. Nevertheless, they set to work without much protest."

The results obtained on 26 observers with the new procedure are reproduced in Fig. 1.4 in the form of medians of 52 estimates and of interquartile ranges. Every observer made two loudness estimates of each stimulus. Because every observer used a different standard, calculation of group medians posed a challenge. The data had to be normalized. This was done by first calculating the grand median of all the data, then the median for each individual observer. The ratios between the grand median and the individual medians provided the normalization factors by which the individual data were multiplied. Since no further normalization occurred, the data of Fig. 1.4 represent the actual median values given by the observers. Significantly, the SPL of 45 dB was associated with the numerical value of unity. Much before

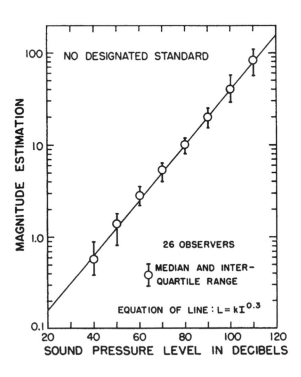

Fig. 1.4 Magnitude estimation of loudness of a 1,000-Hz tone without a designated reference standard. The observers were allowed to choose their own reference standards, and the individual data were normalized to a reference standard of 80 dB SPL called 10. Reproduced from Stevens (1956) used with permission of the author and the University of Illinois Press. Copyright 1956 by the Board of Trustees of the University of Illinois

these measurements were performed, Stevens intuitively defined the loudness of a 1,000 Hz tone 40 dB above its threshold, when heard binaurally, as the unit of loudness, calling it one *sone* (Stevens, 1936). Because the binaural threshold of audibility is about 5 dB above the accepted standard SPL for the 1,000 Hz tone, the coincidence is almost perfect. These relationships are discussed in a different context below.

Stevens seems to have been highly satisfied with the results shown in Fig. 1.4. He wrote: "Those data defined the most satisfactory power function for loudness that had yet been produced" (Stevens, 1956). However, their significance is not clear because of the ambiguity of the instructions. It is clear that Stevens let his observers choose freely the numbers to be associated with the first stimulus presented to them. He seemed to imply that this stimulus was to serve as a reference standard. But what did he mean by telling the observers "to make the ratios between the numbers... correspond to the ratios between the loudnesses of the tones." From the preceding text, we could conclude that he meant numerical loudness ratios with the first stimulus heard. But is this what the observers understood? The ambiguous instructions were perpetuated by Stevens' coworkers. They are still found in Marks's "Sensory Processes" (1974) in connection with an experiment on heat sensation:

> I am going to present a number of heat stimuli to your forehead. Your task is to judge how warm each stimulus feels to you by assigning numbers to stand for the degree of apparent warmth. To the first stimulus, assign whatever number seems to you the most appropriate to represent the degree of warmth. Then, for succeeding stimuli, assign other numbers in proportion of warmth. If one stimulus seems three times as warm as another, assign a number three times as great; if it feels one fifth as warm, assign a number one-fifth as great. Any type of number – whole number, decimal, or fraction – may be used.

In spite of the ambiguity, the instructions clearly refer to ratio estimation, and Stevens' comment – "This final experiment seems to go about as far as an experiment can go toward getting the O's to make absolute judgments of loudness." does not seem entirely justified. Nevertheless, it predicted further fundamental discoveries to come. The germinal ones occurred in connection with the master theses of one of my past students, Hellman (1960). The intended subject of the thesis was innocuous, concerning the shape of the loudness function near the threshold of audibility, but went astray. Stevens tended to use SPLs at and above 40 dB relative to the reference at $0.002 \, \text{dyne/cm}^2$. At lower SPLs, averaging the data becomes problematic because of the intersubject variation of the threshold of audibility. Is it better to refer the data to the $0.002 \, \text{dyne/cm}^2$ sound pressure or to the thresholds of individual observers? To answer this question, Hellman used both reference standards.

All the experiments were performed at the sound frequency of 1,000 Hz. Stevens' original method was used. The observers had a double throw switch at their disposal and were able to turn on either the standard or the variable tone. They could listen to either as often as they wished before making a judgment. A reference standard was chosen either in terms of the SPL or the sensation level (SL) relative to the observer's threshold of audibility. The latter was determined by the method of limits before every series of loudness estimations.

For the first series, an 80 dB standard associated with the number 10 was chosen. Two separate groups of inexperienced observers participated. One consisting of 15 graduate and undergraduate students received the standard in terms of SPL, the other, consisting of a similar population of 14 observers, in terms of SL. Every stimulus was presented twice in a random order that differed from observer to observer. They were presented at 9 SPLs or SLs, keeping the same inter-stimulus ratios in both instances. The results are shown by group medians in Fig. 1.5. The circles referring to the SPL series approximate closely a power function, as was expected

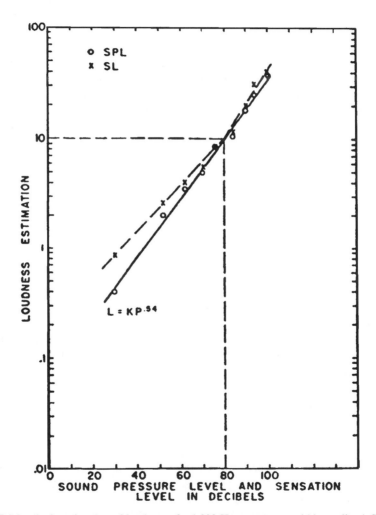

Fig. 1.5 Magnitude estimation of loudness of a 1,000-Hz tone presented binaurally. A SPL or a SL of 80 dB, called 10, served as a reference standard. The circles show medians of two estimates by each of 15 observers for the SPL standard, the crosses do the same for 14 observers for the SL standard. The straight lines are fitted to the data by eye. Reproduced from Hellman and Zwislocki (1961) with permission from the American Institute of Physics

on the basis of Stevens and his associates' publications. However, the results of the SL series, indicated by the crosses, clearly depart from such a function. They follow a shallower slope below the standard and a steeper one above it. The phenomenon appeared puzzling and led us to an extensive investigation.

First of all, we repeated the experiment twice. The first replication was identical to the original experiment and was performed on the same groups of observers. The only difference was that the observers had acquired some experience in loudness estimation. The obtained results did not differ significantly from those of the original experiment. In a second replication, two overlapping groups of 29 and 25 observers, respectively, participated. All were experienced in loudness estimation. Their median threshold of audibility was at 8 dB SPL. Again, the patterns of the SPL and SL functions of Fig. 1.5 were confirmed. As a consequence, we had to accept the difference between the two functions as being systematic, rather than fortuitous.

Analysis of the median data revealed only one factor that could account for the difference between the SPL and SL patterns. Since the median threshold of the observers was at 8 dB SPL, the SPL standard was at a SL 8 dB lower than the SL standard, both associated with the number 10. To find out if such a small difference in the standard SL could produce the divergence between the SPL and SL functions, we divided the observers into two subgroups, one having thresholds between 2 and 6 dB SPL and the other, between 8 and 14 dB. Recomputation of the previously obtained SPL data led to the results shown in Fig. 1.6. The difference in the loudness estimates of the two subgroups, although small, is systematic and clearly apparent. The subgroup with the lower thresholds produced a pattern more like the SL pattern of Fig. 1.5, the subgroup with the higher thresholds, a pattern more like the SPL pattern of that figure. When, instead of a SPL standard, a SL standard was used, the difference disappeared. This means that the observers referred their loudness estimates to SL that, unlike the SPL reference, was independent of the threshold of audibility. Apparently, individual differences in the threshold produced parallel shifts in the loudness function, not affecting its shape. Accordingly, we referred the loudness scales to SL rather than to SPL in subsequent experiments.

First, we investigated the effect of standard SL, keeping the numerical modulus at 10. A group of 9 experienced listeners with a median threshold of 6 dB SPL participated. The standard SLs were placed at 40, 60, 70 and 90 dB. Loudness estimates with the standard SL at 80 dB were obtained previously. The experimental results are shown in Fig. 1.7. The effect of the standard SL is spectacular. Below the standard, the curves become strongly flatter as the standard increases and converge at low SLs. Such a result would not be expected if the observers judged loudness ratios between the standard and the variable stimuli, as they were requested to do. The ratios should not be affected by the absolute value of the standard and would have to produce parallel curves. This is approximately true above the standard were the observers seem to have been able to follow the instructions. At these levels, they had to respond to multiples of number 10, which may be an easier task than estimation of its fractional values.

Stevens (1956), who had obtained similar deviations from pure ratio judgments when he had varied the SPL while holding the standard number constant, had

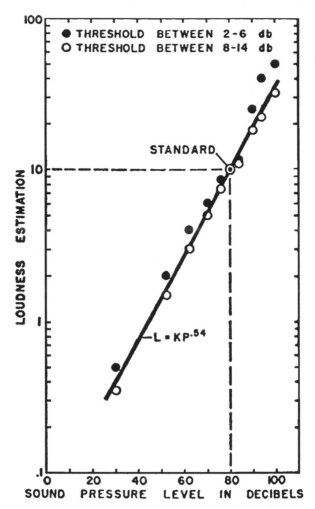

Fig. 1.6 Magnitude estimation of loudness of a 1,000-Hz tone by two groups of observers, one having thresholds of audibility between 2 and 6 dB SPL, the other, between 8 and 14 dB. The estimates were referred to a standard SPL of 80 dB assigned the number 10. The data points refer to median estimates, as in Fig. 1.5. Reproduced from Hellman and Zwislocki (1961) with permission from the American Institute of Physics

suggested that, when the standard sound is very loud or very soft, the observers give ambiguous responses, biasing their ratio judgments by attempts at telling the experimenter that the sound is very loud or very soft in absolute terms. Stevens' observation proved prophetic.

Significantly, the curve of Fig. 1.7 belonging to the standard SL of 70 dB coupled to the standard number 10 has approximately the same slope over its entire course, except for a perturbation near the standard. It is the only curve exhibiting such a characteristic. The constancy is consistent with Stevens' curve of Fig. 1.3

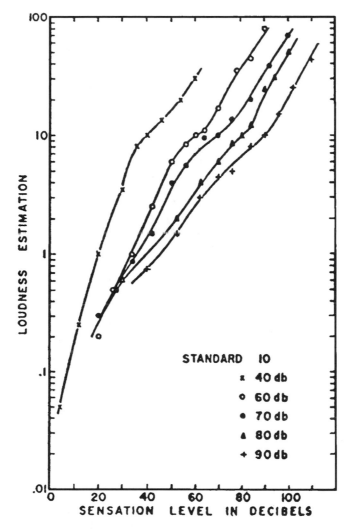

Fig. 1.7 Median loudness estimates by 9 experienced observers relative to 5 standard SLs associated with a modulus of 10. Reproduced from Hellman and Zwislocki (1961) with permission from the American Institute of Physics

associated with a standard SPL of 80 dB and the same standard number. A SPL of 80 dB coincides approximately with an average SL of 70 dB.

To test the symmetry of the relationship between standard numbers and SLs, we performed a complementary experiment in which the standard SL was kept constant and the standard number varied. For the former, we chose a SL of 40 dB for the latter the numbers 0.1, 1, 10 and 100, respectively. A group of 9 sophisticated observers participated in the experiment that was run in the same way as the experiment that produced the results of Fig. 1.7. The data, normalized to the standard number 1, are

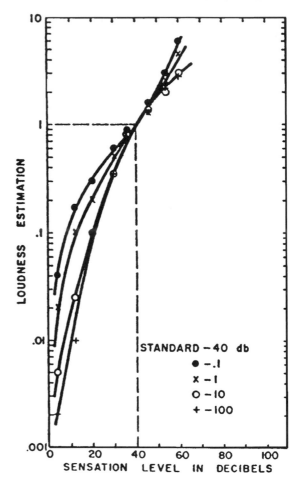

Fig. 1.8 Median loudness estimates by the same 9 observers as in Fig. 1.7 relative to a standard SL of 40 dB associated consecutively with the moduli of 0.1, 1, 10, and 100. Reproduced from Hellman and Zwislocki (1961) with permission from the American Institute of Physics

shown in Fig. 1.8. The dependence of the curve slope on the relationship between the standard SL and the standard number is evident. The slope increases below the standard as the standard number is increased; above the standard, the relationship is reversed. The slope variation is consistent with that of Fig. 1.7. Below the standard, the slope increases as the numerical ratio between the standard SL and the standard number decreases; above the standard, the slope decreases. For the standard pair: 40 dB SL and 1, the slope below and above the standard is the same, except at very low SLs where it increases for all the curves. The latter phenomenon is quite general for sensory magnitude functions. Its possibly fundamental nature is discussed in the next chapter.

The curve of Fig. 1.8 preserving the same slope on both sides of the standard coincides roughly with the curve having an approximately constant slope below and above the standard in Fig. 1.7, which has been generated by the standard pair: 70 dB SL and number 10. An even better coincidence is obtained when the ordinates of

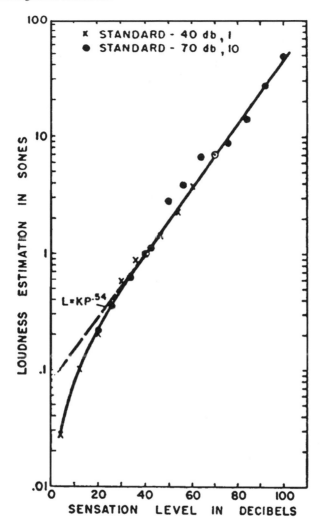

Fig. 1.9 Composite of the data obtained with two reference standards: 40 dB SL called 1, and 70 dB called 10. The latter data points were shifted vertically according to a multiplier of 0.7. The curve, drawn by hand, approximates both sets of data. Above 30 dB, it follows a power function with an exponent of 0.54 (0.27 in energy terms). Reproduced from Hellman and Zwislocki (1961) with permission from the American Institute of Physics

the latter curve are multiplied by 0.7. Evidently, number 7 would have been more appropriate for the SL of 10 dB. When the correction is made, both curves come to practically complete coincidence, as is shown in Fig. 1.9. The individual data points obtained with either standard show only insignificant departures from a common curve, except in the vicinity of the standards. The perturbations occurring there, especially near the 70 dB, number 7 standard, are probably due to the constraint introduced by the imposed standard and its repeated presentation.

The results synthesized in Fig. 1.9 suggest that an invariant function is obtained only for particular standard pairs of SLs and numbers. Accordingly, the coupling between loudness and the number continuum appears to be absolute rather than relative, so that invariant loudness functions are based on absolute rather than relative units. Further supporting evidence for this conclusion can be found in Stevens' experiment in which observers were allowed to choose their own reference standards and which he characterized as an experiment that went as far "as an experiment can go toward getting the Os to make absolute judgments of loudness" (Stevens, 1956). His results are compared to the loudness curve of Fig. 1.9 in Fig. 1.10. For this purpose, the curve plotted in Fig. 1.9 in terms of SL, had to be replotted in terms of SPL. It has been moved to the right by 6 dB, the median threshold of audibility of the participating observers. The adjustment has produced an almost complete agreement with Stevens' data in absolute terms. It suggests that both, the two standard SLs and the numbers on which the curve is based, as well as Stevens' results, are consistent with absolute numerical loudness estimates. The curve approximates a power function, except at low SPLs where its slope is increased.

An additional example of what happens when the reference standards do not agree with the inferred absolute units is provided by an experiment on 16 observers, in which Stevens (1956) determined two loudness functions, one based on a standard pair of 30 dB SPL and 1, the other, on a standard pair of 120 dB SPL and 100. The

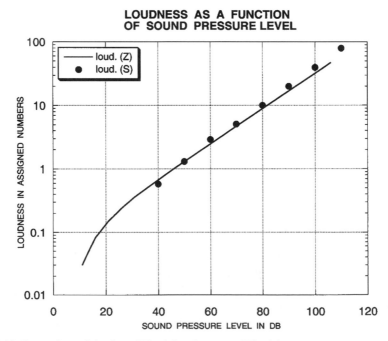

Fig. 1.10 Comparison of the data of Fig. 1.4 to the curve of Fig. 1.9

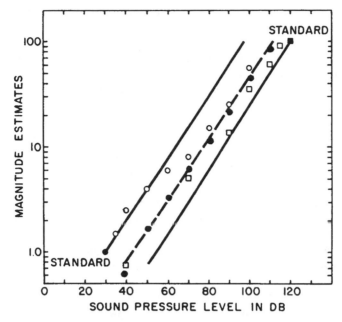

Fig. 1.11 Median magnitude estimates of loudness of a 1,000-Hz tone by 16 observers relative to two consecutive standards 30 dB SPL, 1 (*circles*) and 120 dB SPL, 100 (*squares*). The straight lines, drawn through the two standard points follow power functions with the exponent 0.3. The parallel dashed line joins the top data points obtained with the first standard to the bottom data points obtained with the second standard. The data points marked by filled circles were obtained without a designated standard and are taken from Fig. 1.4 (Data points from Stevens, 1956). Reproduced from Stevens (1956) from the *American Journal of Psychology*. Copyright 1956 by the Board of Trustees of the University of Illinois. Used with permission of the author and the University of Illinois Press

obtained data follow power functions with the expected exponent of about 0.3 near the standards but gradually converge on an implied intermediate curve at stimulus values that are more distant. The intermediate curve was shown in a past article to coincide approximately with the absolute loudness function of Fig. 1.10 (Zwislocki and Goodman, 1980). The corresponding graph is reproduced in Fig. 1.11 with the addition of Stevens' loudness data of Fig. 1.10 (*filled circles*).

The relationships of Fig. 1.11, together with those of Fig. 1.7, suggest that, when designated standards are used, and the observers are instructed to judge loudness ratios relative to these standards, they may do so for small ratios but tend to switch to different, more natural standards, when the ratios become large. Natural standards mean in effect that observers make their judgments according to absolute scales based on their own units that cannot be manipulated easily by the experimenter (Hellman and Zwislocki, 1961). Such units, which may differ from one observer to another, imply that numbers acquire absolute subjective magnitudes, and magnitude estimation consists of matching subjective magnitudes of numbers to subjective magnitudes of loudness (Hellman and Zwislocki, 1963). On the basis of this recognition, the reference standard was abandoned entirely, and a method developed in which the observers were simply requested to match their subjective number magnitudes to those of loudness (Hellman and Zwislocki, 1963, 1964, 1968). The instructions differ fundamentally from the instructions in which the observers are allowed to choose their own reference standards but have to refer their subsequent responses to them, in other words, estimate the ratios between loudness magnitudes of subsequent stimuli and the loudness magnitude of the initial stimulus (Stevens, 1956; Marks, 1974). Estimation of the ratio between two variables appears to require a higher-order mental process than the process of matching the magnitudes of two variables. Accordingly, it is possible that observers do not strictly obey the ratio-estimation instructions and replace ratio estimation by the easier process of matching.

Eventually, the method of matching subjective number magnitudes to loudness magnitudes has become generalized to other sensation magnitudes and formalized under the name of "absolute magnitude estimation" (AME) (Zwislocki and Goodman, 1980). By contrast to ratio-estimation procedures, which are limited to the determination of the rate of growth of sensation magnitudes, it has the great advantage of preserving their absolute values. This recognition has profound consequences and is discussed extensively further below.

Caveats concerning the absolute subjective magnitudes of numbers have to be introduced, however. The coupling between these magnitudes and the numbers is not entirely rigid. It varies within a range of about 1–10 from one individual to another and depends to some extent on the details of the scaling procedure. Fractions pose a particular problem because they do not fit properly the concept of numerosity from which the absolute values of numbers are undoubtedly derived. There is some reluctance on the part of the observers to use them. Through experience, methods have been developed that minimize the resulting biases. For example, observers estimate the same magnitudes presented in random order three times. The third estimate usually agrees with the second but the numbers used in the first estimate tend to be larger (e.g. Hellman and Zwislocki, 1963). Why this is so is not entirely clear. Possibly, the observers want to keep the numbers large enough to avoid the need for fractions. The averaged numbers used in the second and third estimates tend to be stable, however, and show an amazing invariance among groups of 5 or more observers. The stability is documented in a further section of this chapter.

If numbers acquire absolute magnitudes, the curve patterns of Figs. 1.7 and 1.8 obtained with designated reference standards can be accounted for by a set of mathematical inequalities. Given subjective loudness magnitudes, L, and subjective magnitudes of numbers, Ψ_N, and the assumption that, in AME, the observers tend to equate the two magnitudes, we have

$$\Psi_N = L \tag{1.4}$$

If, on the other hand, a numerical standard, N_S, is imposed such that its subjective magnitude, Ψ_{NS}, is greater than that of the imposed loudness standard, L_S, $(\Psi_{NS} > L_S)$ the ratios between the subjective variable- and standard-number magnitudes should tend to be smaller than the ratios between the corresponding loudness magnitudes,

$$\frac{\Psi_N}{\Psi_{NS}} < \frac{L}{L_S} \tag{1.5}$$

For easier interpretation, this inequality can be rewritten as

$$\frac{\Psi_{NS}}{L_S} > \frac{\Psi_N}{L} \tag{1.6}$$

On the assumption that the observers tend to preserve the imposed standard ratio near the standard, so that $\Psi_N/L = \Psi_{NS}/L_S > 1$, and tend to match the subjective number and loudness magnitudes far away from the standard, so that $\Psi_N/L \to 1 < \Psi_{NS}/L_S$, the loudness curve should become steeper below the standard and flatter above it. The reverse should be true if the imposed standard number were too small relative to the standard loudness, so that its subjective magnitude $\Psi_{NS} < L_S$.

The above inequalities imply that, when N_S is chosen so that $\Psi_{NS} = L_S$,

$$\frac{\Psi_N}{\Psi_{NS}} = \frac{L}{L_S} \tag{1.7}$$

the slope of the curve remains invariant below and above the standard. When the curve follows a power function, it can be described with the help of Eq. (1.4) by the logarithmic relationship

$$\log \Psi_N = \log L = C \log \left(\frac{P}{P_0}\right) \tag{1.8}$$

where C is a dimensional constant, P – a sound pressure amplitude, and P_0 – a reference sound pressure amplitude. If the assumption is made that the subjective magnitudes of numbers are directly proportional to the numbers themselves, and the expression $\log (P/P_0)$ is replaced by the expression for SPL, or SL, the latter equation can be rewritten in the form

$$\log N_L = k \, \theta \, 20 \log \left(\frac{P}{P_0}\right) \tag{1.9}$$

in which N_L stands for the numerical estimate of loudness and k is another dimensional constant. Note that the equation involves two crucial assumptions – first, that numbers acquire absolute subjective magnitudes, second, that these magnitudes are matched to the loudness magnitudes. These assumptions are further justified below.

The conclusion that, in magnitude-estimation tasks, observers tend to make absolute judgments of sensation magnitudes is supported by many experiments in addition to the ones described above. Two examples are given next.

In the first (Zwislocki and Goodman, 1980), 12 observers estimated the loudness of repeated 20-msec bursts of a 1,000-Hz tone presented monaurally by means of special insert phones that reduced the residual noise present in the sound proofed booth the observers were placed in. All the observers had normal hearing according to audiometric tests. They could neither see nor hear the experimenter, except through an intercom system. In every experimental session, the stimuli were presented in three sequences, called runs, every one of which contained a full complement of the stimuli of the session. As a consequence, every stimulus was judged three times. The results of the second and third runs were averaged.

The tone-bursts were presented at various SLs in a random order that differed from observer to observer, run to run, and session to session. The observers were allowed to listen to the burst sequences as long as they wished before assigning a number. Nevertheless, they were urged to respond as spontaneously and promptly as they were able to. After a number was assigned, the tone bursts were interrupted for a short time period before being restarted. Prior to the experiments, every observer was shown several lines of varying length drawn on a blackboard, one at a time, and asked to assign numbers to them according to their lengths. The line sequence started with a medium line length, and some very short lines were included to force the observers to use fractions. The experimenter explained to the observers that there can always be a shorter or longer line than the lines contained in the sequence they saw. He/she also explained that similar relationships existed for the numbers, and that it was not possible to run out of them at either end. Previous experience showed that informal scaling of line lengths stabilized subsequent scaling of other continua. This may be so because the line length appears to present a well defined psychological magnitude.

After the preliminary line-length scaling, the observers were informed that they will hear repeated tone bursts and that their task consisted of assigning numbers to them in such a way that the subjective magnitudes of the numbers matched the loudness magnitudes of the tone bursts. They were told to concentrate on the tone bursts they were hearing and not to be concerned about the numbers they had associated with the loudness magnitudes of preceding tone bursts. The experimenter explained to them further that they could use any positive numbers that appeared appropriate to them – whole numbers, fractions or decimals. They should not think of any rules they might have learned previously about number assignment and be as spontaneous and prompt in their responses as possible.

The observers were divided into two groups of 6. One group received the tone bursts within the SL range of 6–54 dB, the other, within the range of 30–78 dB. The results obtained in the first run of the first session, before the observers had any prior experience with magnitude estimation, are discussed first. If the observers matched subjective number magnitudes to loudness magnitudes on similar absolute scales varying randomly among the observers, the median assigned numbers should have fallen on the same curve for both groups and should have coincided within the region of overlap of the two SL ranges. The graph of Fig. 1.12 indicates that the obtained data satisfied approximately both requirements. The data points of both ranges fell roughly on one and the same curve. The implied curve is consistent with a power function except at low SLs where it bends downward, in agreement with loudness curves obtained previously and mentioned above. In the region of overlap, the data points belonging to the group receiving the higher range and indicated by the unfilled circles lie only slightly below the data points of the group receiving the lower range (*filled circles*), and both sets remain within one standard error and within the differences usually found between groups of this size. The difference

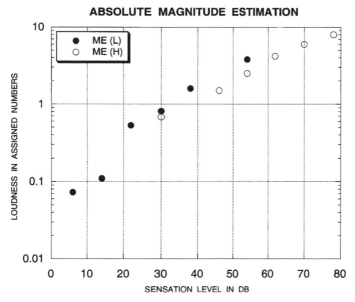

Fig. 1.12 Absolute magnitude estimates (AME) of loudness at 1,000 Hz by two experienced groups of 6 observers, each receiving a different range of SLs – 6–54 (*filled circles*) and 30–78 dB (*unfilled circles*). The ordinate differences between the data points of the two ranges within their overlap are likely to stem from the group differences rather than from a range effect (see text). Data from Zwislocki and Goodman (1980)

between the two groups remained about the same in the second and third runs of the session after the observers had the opportunity to become familiar with the loudness ranges involved. It was not materially altered even after the ranges had been exchanged between the two groups, as is documented in Fig. 1.13. Because one group gave slightly greater numerical estimates than the other group, irrespective of which sequence of SL ranges the groups were exposed to, the persistent difference must be ascribed to an inherent difference between the two groups and not to a difference in range location. The almost complete absence of an effect of the range location was confirmed in later studies of AME (e.g. Gescheider and Hughson, 1991). The effect of the location was clearly evident with the ME instructions, however, according to which the observers choose the reference standard but must refer to it in their responses to subsequent stimuli (e.g. Marks, 1991). The almost complete absence of the effect of range location with AME instructions leaves little doubt that, at least with these instructions, the numbers are assigned to loudness magnitudes according to absolute units. The same units appear to have underlied all the loudness functions obtained without a designated reference standard and most of those referred to a standard whose coordinates coincided with these functions, as exemplified in Figs. 1.3, 1.4, and 1.9–1.13. At a constant sound frequency of 1,000 Hz, they followed roughly the same function that goes approximately through the point: 40 dB SL, 1, in agreement with the sone unit chosen intuitively by Stevens

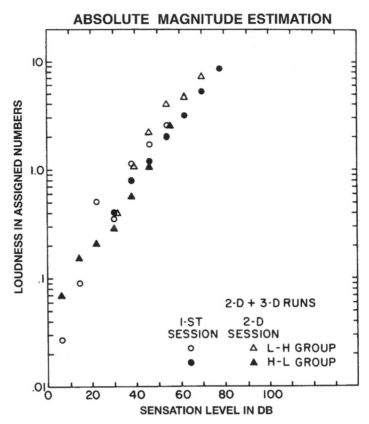

Fig. 1.13 The effect of experience on AME of loudness of a 1,000-Hz tone by the two groups of observers of Fig. 1.12 after the SL ranges had been switched between the groups. The circles belong to the first estimates (*first run*) of the first session, the triangles to an average of the second and third runs of the second session. Note that even after the range reversal, the same group tended to produce the higher magnitude estimates. Data from Zwislocki and Goodman (1980) reproduced with permission from the Psychonomic Society

many years before the method of magnitude estimation was introduced (Stevens, 1936). The function obeys Stevens' power law, except at low SLs where it becomes gradually steeper.

The second example bringing absolute scaling into evidence concerns line length rather than loudness (Zwislocki and Goodman, 1980). This time, the effect of the number range was investigated. Two groups of observers were involved – a group of 6 adults whose number range was practically unlimited and a group of 12 children 5–6 years of age, whose number range was confined within 1 and 99. The children did not know fractions or numbers greater than 99. The stimuli consisted of 7 black lines projected on a reflecting screen, one at a time, in a random order of length, except that neither the shortest nor the longest lines were presented first. The line length ranged from 1 to 100 cm, and the lines were viewed by the adults at a distance of 3 m, and by the children at that of 3.7 m. Because of the phenomenon of size constancy, the moderate distance difference was immaterial (e.g. M. Teghtsoonian and Beckwith, 1976). Neither the adults nor the children had any experience with magnitude estimation. The children were subdivided in 2 equal subgroups of 6, one receiving a line of 53.5 cm the other, a line of 2.8 cm as the first stimulus. The adults assigned fractional numbers to the latter. The adults made only one estimate of subjective length per line because they remembered the assigned numbers too well for a second estimate to be meaningful. The children made two estimates, but only the first was used for comparison with the estimates made by the adults. Quoting from Zwislocki and Goodman (1980),

> The adults were simply instructed to assign numbers to lines so that the subjective magnitudes of numbers matched the subjective magnitudes of line lengths. They were asked to make intuitive responses and not to think of physical measures of length or rules of arithmetic. The children were introduced to the scaling procedure by the following set of questions to which they could respond 'yes' or 'no'. 'Do you know some large numbers? How about some small numbers? Can you think of numbers that are neither small nor large? Can you imagine a line that is long? Can you imagine a short line? How about a line that is neither very long nor very short?' Then they were told: 'Now, I am going to show you lines of various lengths on the screen. Do you think you can tell me which numbers fit the lines you will see? O.K. Let's try it. Here is the first line.

The geometric means of the resulting data are shown in Fig. 1.14. Crosses refer to the responses of the adults. Filled circles refer to the responses of the children who first received the 2.8-cm line length, the unfilled circles, to the responses obtained with the 53.5-cm line length as the first stimulus. The lines were fitted to the data by eye, and the brackets indicate the standard deviations of log scores relative to the geometric means of either group. Clearly, the data of both subgroups of the children agree with each other and with those of the adults within the range of the numbers known to the children. The limited range of the numbers does not appear to have affected either the position or the slope of the curve fitting the children's data. In particular, the children did not adjust their numerical responses to avoid fractions but, instead, truncated the numbers at 1. When very short lines were presented to them, some children hesitated whether to associate with them a numerical value of 1 or 0. They invariably decided on 1 in the end, probably, because a line was present after all, and 0 would have meant no line. The results obtained on the children were amply confirmed by Collins and Gescheider (1989) and extended to loudness. They also agree with the results of Teghtsoonian and Beckwith (1976) who asked children having the ages of 8, 10, 12 and 18 years to scale the height of rectangles presented at various distances. No significant age effect was found. The age independence was further extended by Verrillo (1981) to 68 years.

The children's responses documented in Fig. 1.14 and confirmed in the articles cited above would be difficult to explain without assuming that numbers had for them absolute rather than relative values. These values were most likely acquired during the learning process of the number system. Children learn numbers by counting objects rather than in the abstraction. Based on the agreement between the responses of the children and of the adults, the acquired values tend to be preserved throughout the lifetime. Interestingly, some inter-individual differences creep in and

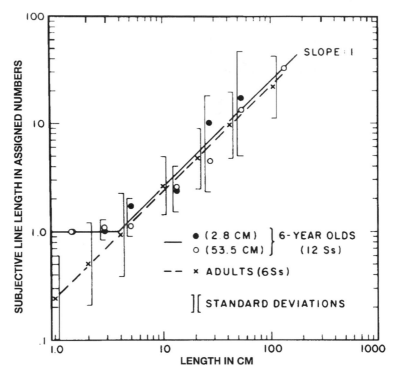

Fig. 1.14 AME of line length by 6 adults (*single magnitude estimates*) and by two groups of 6 children 5–6 years of age, one receiving a 2.8-cm line first the other starting with a 53.5-cm line (*first magnitude estimates*). The lines were fitted to the data by eye. Reproduced from Zwislocki and Goodman (1980) with permission from the Psychonomic Society

are present already at an early age, as suggested by the rather large standard deviations of the children's responses. They may depend in part on the size of the objects children learn to count. A given number of pencils or small pebbles may appear psychologically smaller than the same number of oranges or apples. In any event, psychological magnitudes of numbers appear to be rooted in numerosity whose units are absolute.

Note that the numbers associated with variable line lengths do not agree with any established physical measures, like centimeters or inches, so that neither the adults nor the children attempted to estimate the physical line length. Their numerical estimates must have reflected the subjective length magnitudes they experienced. The absolute nature of these estimates is further underscored by the fact that line-length functions obtained by different investigators without the use of a designated reference standards for horizontal lines all go through approximately the same point of 4 cm, 1 (e.g. Verrillo, 1979a, 1983; Zwislocki and Goodman, 1980; Zwislocki, 1983; Collins and Gescheider 1989).

The numerical estimates of subjective line-length magnitudes follow approximately straight lines with a slope on the order of 1 on double-log coordinates of

Fig. 1.14. In other words, they obey power functions with the exponent of about one, or direct proportionality. Therefore, on the implicit assumption that the subjective magnitudes of numbers are directly proportional to the numbers, the subjective line length grows in direct proportion to the physical line length. An explicit justification of the assumption is given in the section on the validity of the power law.

1.3 Magnitude Production and Magnitude Balance

In magnitude estimation, numbers are assigned to psychological magnitudes. According to the preceding analysis, the observers do so predominantly by matching subjective magnitudes of numbers to other subjective magnitudes, such as those of loudness or brightness. If subjective magnitudes of numbers can be matched to other subjective magnitudes, an inverse operation should be possible in which the latter are matched to the former. The observer is given a number and is instructed to match to its subjective magnitude the subjective magnitude of another variable belonging to a specified modality. Stevens (1958) proposed such a procedure to counterbalance biases he suspected to be present in magnitude estimation – an asymmetric, unidirectional operation, and called it magnitude production. If both magnitude estimation and production consist of matching psychological magnitudes of numbers and another psychological variable to each other, their names may appear as somewhat inappropriate. Nevertheless both names have been widely accepted, and I shall use them as somewhat illogical but convenient designations. Languages are full of these. In application to absolute scaling, magnitude production received the name of absolute magnitude production (AMP) as a counterpart to absolute magnitude estimation (AME) (Zwislocki and Goodman, 1980).

When Stevens and Guirao, (1962) applied magnitude estimation and magnitude production to loudness, they saw that, indeed, they did not produce identical results, as can be seen in Fig. 1.15. When sound intensity was plotted on the abscissa axis and the numbers associated with loudness on the ordinate axis, magnitude estimation produced a shallower curve slope than magnitude production. Stevens' original assumption was correct – the difference between the two curves suggests that magnitude estimation is biased. However, the same may apply to magnitude production so that both can be expected to give biased results. The relative bias between the two procedures was confirmed on many occasions (e.g. Stevens and Greenbaum, 1966). Stevens thought that the bias could be due to what he called a "regression effect" – the tendency on the part of the observers to shorten the range of stimuli under their control. So, for example, in magnitude estimation of loudness, they would shorten the range of numbers, in magnitude production, the range of sound intensities. Usually, the assumption is made that the biases are opposite and equal and the data of both are averaged by one or another statistical procedure in the hope of obtaining a minimally biased result (e.g. Hellman and Zwislocki, 1963, 1964; Verrillo et al., 1969).

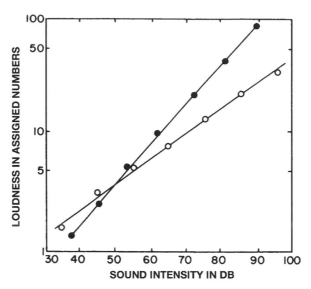

Fig. 1.15 Magnitude estimation (*unfilled circles*) and magnitude production (*filled circles*) of loudness of a 1,000-Hz tone. Every data point results from a geometric average of two responses. Modified from Stevens and Guirao (1962), as reproduced in Geischeider (1997). Permission to reproduce from Lawrence Erlbaum Associates

In absolute scaling, the relative bias between AME and AMP is not limited to the slope of the curves but AME produces slightly higher numerical values than AMP. Nevertheless, one feature remains invariant –whether ratio scaling or absolute scaling is employed, both magnitude estimation and magnitude production generate power functions, except at low stimulus intensities at which the curve slope tends to increase.

Measurements of loudness performed by Hellman and myself can serve as an example of a combined AME – AMP procedure that we called "numerical magnitude balance" (e.g. Hellman and Zwislocki, 1963, 1964, 1968). The observers first made absolute magnitude estimates of loudness. Subsequently, the range of numbers they covered was used as a basis for the numbers included in absolute magnitude production. In each procedure the variable under the experimenter's control was presented in three runs in a random order of magnitudes, which differed among the runs. The data obtained in the last two runs were averaged to produce geometric means that were interpolated by smooth curves. The data of the first run, which usually deviate the most from the remaining data were left out. The AME and AMP curves were geometrically averaged to obtain the mean magnitude- balance curve. We attempted to choose experimental conditions under which direct validation of the magnitude-balance results was possible by direct loudness matches. Perhaps the clearest results were obtained in an experimental series in which a 1,000 Hz tone was masked partially by broadband noise (Hellman and Zwislocki 1964). Such masking elevates the threshold of audibility and increases the slope of the loudness

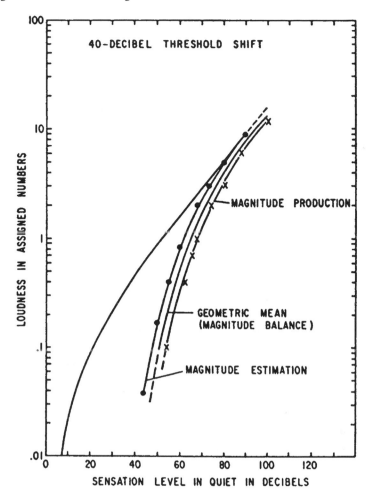

Fig. 1.16 Loudness of a 1,000-Hz ton in the absence (*upper curve*) and presence of a partially masking noise. Data points indicated by circles were obtained by AME, those indicated by crosses, by absolute magnitude production (AMP). Their geometric means are designated as "numerical magnitude balance" data. The upper curve was also determined by the numerical magnitude balance. Reproduced from Hellman and Zwislocki (1964) with permission from the American Institute of Physics

function – a phenomenon called "loudness recruitment." If the noise is presented to one ear the increased slope can be measured relative to the slope in the unmasked ear by direct loudness matching. Theoretically, it should be possible to recover the same slope relationship from the magnitude-balance procedure, if applied to each ear in turn. The results of one such experiment are reproduced in Fig. 1.16. They are plotted over an abscissa axis of SL determined in the absence of masking noise. Introduction of the noise increased the threshold of audibility by 40 dB. The solid curve to the left approximates the magnitude-balance data determined in the unmasked ear in

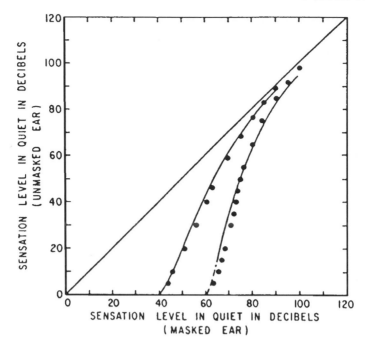

Fig. 1.17 Loudness-level curves obtained by numerical magnitude balance in the presence of two levels of partially masking noise are compared to corresponding direct loudness matches by the method of adjustment (*circles*). Reproduced from Hellman and Zwislocki (1964) with permission from the American Institute of Physics

the presence of masking noise in the contralateral ear. The filled circles indicate the results of AME of loudness in the masked ear, the crosses, the corresponding AMP results. The solid curve between the two data sets follows the numerical magnitude-balance values obtained by calculating the geometric means of the numerical AME and AMP values. Horizontal cuts through the curves of Fig. 1.16 and through corresponding curves obtained for a threshold shift of 60 dB, yielded the smooth curves reproduced in Figs. 1.17 and 1.18. The curve ordinates in Fig. 1.17 indicate the SLs in the unmasked ear required for loudness equality with the masked ear at two masking levels and in the absence of masking, based on the loudness-balance results. The curves in Fig. 1.18 are based on the results of magnitude production alone. In both figures, the filled circles show corresponding results of direct loudness matching between the masked and unmasked ears. Somewhat surprisingly, direct loudness matching is in better agreement with AMP than with magnitude balance. Accordingly, magnitude estimation seems to produce a greater bias than magnitude production. Such asymmetry was already suspected by Stevens and Guirao, (1962).

AMP was also found to be more stable than AME in some later experiments on loudness (Zwislocki and Goodman, 1980). One experiment concerned the effect of the observers' experience. A group of 18 observers, who had never before participated in such procedures, performed the AME; another group of 12 inexperienced observers, the AMP. The observers listened to 20-msec bursts of a 1,000 Hz tone

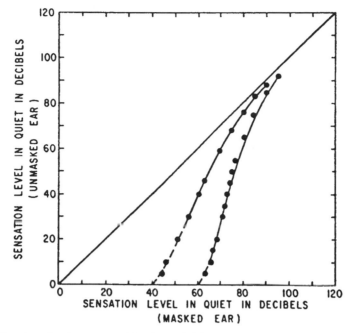

Fig. 1.18 Loudness level curves resulting from AMP are compared to loudness matching data of Fig. 1.17. Reproduced from Hellman and Zwislocki (1964) with permission from the American Institute of Physics

repeated at a rate of one per second. The tone bursts were presented in two sessions comprising 3 runs each. Within every run, the stimuli were presented at seven SLs in random order that differed from run to run. The observers were allowed to listen to the tone bursts at every level until they made their decision. The experimental results are shown in Fig. 1.19 by triangles and circles for the AME and by crosses and asterisks for the AMP. The triangles indicate the AME data obtained in the first run of the first session, before the observers had any chance of learning the loudness range. The unfilled circles indicate the geometric means of the data obtained in the second and third runs of the first session, the filled circles, similarly obtained means in the second session. Note that the assigned numbers decreased gradually with the observers' experience, especially at low SLs. As judged from the individual data, the relatively high values of the first run in the first session may have been due in part to the observers' reluctance to use fractions, which do not belong to the set of whole numbers included in the concept of numerosity. By contrast, experience had little effect on the observers' responses in AMP, in which fractional numbers were included. There is no significant difference between the SL values chosen by the observers in the first run of the first session and those chosen in the subsequent runs, whether in the first or second session. The amazing lack of an effect of experience on the magnitude-production responses can be understood only by assuming that the observers had stored in their minds absolute psychological magnitudes of the

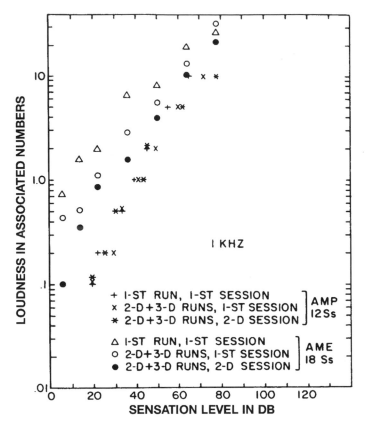

Fig. 1.19 Loudness of 1,000-Hz tone bursts, as measured by the methods of AME (triangles and circles) and AMP (*crosses and stars*) on two groups of observers (*Ss*) in several sessions. Note that the effect of experience on AME is much greater than on AMP. Reproduced from Zwislocki and Goodman (1980) with permission from the Psychonomic Society

numbers given them by the experimenter. This conclusion is in agreement with the magnitude-estimation results of Fig. 1.14 obtained on children who did not know fractions. The greater effect of experience on magnitude-estimation results evident in Fig. 1.19 can be eliminated almost completely by prior magnitude estimation of line lengths, when medium-length lines are shown first, then, lines so short that the observers see themselves forced to use fractions. The absolute loudness estimates obtained in this way and shown in Figs. 1.12 and 1.13 agree approximately with the magnitude production results of Fig. 1.19.

Because of its stability, perhaps AMP should be the method of choice in scaling experiments. However, it is often difficult to implement. Fortunately, biases inherent in AME can be minimized by auxiliary procedures, such as prior scaling of line length whose psychological magnitude appears to be more clearly defined than other psychological magnitudes, such as those of loudness and brightness. Joint measurements of subjective line length and loudness revealed that the inter-observer variability in loudness estimates was highly correlated with the deviations from the exponent of unity in line-length estimates. As a consequence, the variability could be almost completely removed by dividing the measured individual exponents of loudness functions by the deviations from the unity exponent in corresponding line-length functions (Zwislocki, 1983; Collins and Gescheider 1989).

On the whole, when the pitfalls of absolute scaling, in particular of AME, are avoided, the scaling proves to be astonishingly stable with a great deal of generality. A convincing example is provided by two loudness studies performed at two quite different locations, on two different groups of observers, by two different teams of experimenters who were not in communication with each other (Zwislocki and Goodman, 1980). It concerns 1,000-Hz tone bursts presented monaurally. One study was performed on 9 observers the other on 10. In both studies, the observers performed AME as well as AMP procedures. They were instructed to match freely subjective magnitudes of numbers and loudness without being given any reference standard. The results are reproduced in Fig. 1.20. The circles refer to AME, the crosses to AMP. Above 40 dB SL, the results of both studies nearly coincide, and there is no systematic difference between the data obtained with the two procedures. The divergence is more substantial at lower SL, where fractional numbers come

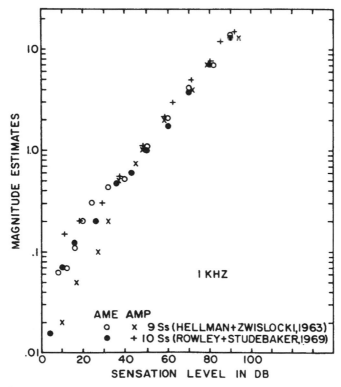

Fig. 1.20 AME and AMP of 1,000-Hz tone bursts. Comparison of the results obtained on two different groups of observers in two different laboratories by different researchers. Reproduced from Zwislocki and Goodman (1980) with permission from the Psychonomic Society

into play. Somewhat surprisingly, this is particularly true for AMP. Geometrically averaging the data of the two procedures to obtain numerical-balance values would nearly eliminate the divergence. The agreement between the two studies strengthens the case for the stability and generality of the implied subjective units.

1.4 Cross-Modal Matching and Transitivity

In cross-modal, or cross-modality matching, the sensation magnitude of one modality is matched to that of another modality. For example, the loudness of a sound is matched to the brightness of a light. To achieve a match, the observer varies the stimulus magnitude under his control. The results are recorded as numerical values of this magnitude versus the numerical values of the stimulus magnitude given by the experimenter. In a counterbalanced procedure, first one of the two stimuli is placed under the observer's control, then, the other. The results of the first experiment on cross-modal matching are reproduced from Stevens (1975) in Fig. 1.21. They concern tactile vibration and acoustic noise whose subjective magnitudes were matched in the counterbalanced procedure. Note that the obtained data follow power functions and that they are subject to the regression effect, just like in magnitude estimation and production.

The experiment was performed in 1958 to answer criticisms directed at the methods of magnitude estimation and production (Stevens, 1959, 1975). Some psychologists, among them Clarence H. Graham, a prominent psychophysicist,

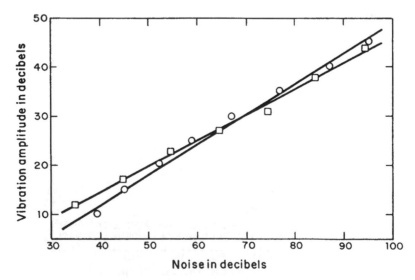

Fig. 1.21 The results of the first experiment on crossmodal matching performed in 1958. The subjective magnitudes of tactile vibration and acoustic noise were matched in their subjective magnitudes as functions of their physical intensities. Reproduced from Stevens (1975) with permission from John Wiley & Sons, Inc.

questioned if subjective coupling of numbers to sensation magnitudes constituted measurement. The success of the experiment demonstrated that people are able to match consistently sensation magnitudes of one modality to sensation magnitudes of another modality. If they can do this without an intermediary of numbers, as demonstrated in the next section, the discovery may be as significant as that of the power law itself. Stevens commented on his pioneering experiment in these words: "Since an 'experimentum crucis' seemed to be involved, it was not without a certain trepidation that I set up the equipment for that test in the spring of 1958." The discovery implies that observers are able to abstract from sensations having diverse qualities their magnitudes that constitute a feature they all have in common. As should be self-evident and is discussed more extensively in the next section, only common features can be compared quantitatively.

Once the feasibility of cross-modal matching had been demonstrated, the procedure was applied to many sensory modalities. The force of handgrip that was measured by means of a calibrated hand dynamometer was one of the preferred reference modalities (Stevens et al., 1960). Squeezing the dynamometer produced a sensation of strain whose subjective magnitude was compared to the subjective magnitudes of brightness, loudness, vibration, pressure, cold, warmth, lifted weights and even electric shock. Subjective magnitudes of numbers were also included in an apparently unintended anticipation of absolute scaling based on the assumption that numbers acquire absolute psychological magnitudes (e.g. Hellman and Zwislocki, 1963; Zwislocki and Goodman, 1980). The results of the dynamometer experiments are reproduced in Fig. 1.22. The force of handgrip is plotted as a function of the

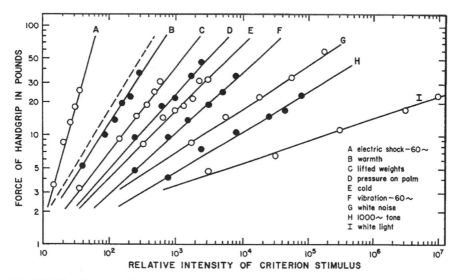

Fig. 1.22 Matching the subjective magnitude of the force of handgrip to subjective magnitudes of the stimuli of several modalities. The required physical force of the handgrip is plotted over the physical magnitudes of the other stimuli referred to relative scales. Reprinted from Stevens (1960) with permission from Blackwell Publishing

relative stimulus intensity for every one of the sensation modalities involved. All the resulting functions obey the power law. Their power exponents differ from those obtained by magnitude estimation and production because the reference function of muscular strain, as measured by magnitude production, has an exponent of 1.7 rather than unity (Stevens and Mack, 1959).

Cross-modal matching makes it possible to verify the mutual consistency of the numerical-estimation results. If numerical scaling generates the power function

$$\psi_1 = k_1 \phi_1^\alpha \tag{1.10}$$

in one sense modality and

$$\psi_2 = k\phi_2^\beta \tag{1.11}$$

in another, equating ψ_1 and ψ_2 produces the theoretical cross-modal match

$$k_2 \phi_2^\beta = k\phi_1^\alpha \tag{1.12}$$

that leads to the predicted intermodal power function

$$\phi_2 = \left(\frac{k_1}{k_2}\right)^{1/\beta} \phi_1^{\alpha/\beta} \tag{1.13}$$

with an exponent equaling the ratio of the exponents obtained in numerical scaling. The predicted function can be verified by a direct experiment in which the two magnitudes, ψ_1 and ψ_2, are matched. If β is the exponent of the power function relating the subjective impression of muscular strain associated with the force on the dynamometer and α is the exponent of the power function relating another subjective magnitude to the magnitude of the physical stimulus that evokes it, then, $\tilde{\alpha}/\beta$ can be calculated. Table 1.1 shows the results of such an operation and compares them to the exponents obtained directly by cross-modal matching (Stevens, 1975).

The approximate agreement between the empirical and predicted exponent values demonstrates the mutual consistency of numerical ratio scaling and cross-modal matching with respect to the exponent. A similar result was obtained with loudness as the reference modality (Stevens, 1966).

Table 1.1 Power exponents obtained by cross-modal matching with the force of handgrip and derived from corresponding magnitude estimations

	Exponent obtained by handgrip	Predicted exponent
Electric shock	2.13	2.06
Warmth on arm	0.96	0.94
Heaviness of lifted weights	0.79	0.85
Pressure on palm	0.67	0.65
Cold on arm	0.60	0.59
Vibration (60 Hz)	0.56	0.56
Loudness of white noise	0.41	0.39
Loudness of 1-kHz tone	0.35	0.39
Brightness of white light	0.21	0.20

The fact that cross-modal matching produced power functions was regarded initially as a validation of the power law. Several investigators pointed out, however, that the proof contained a tacit unproven assumption, namely, that numbers are directly proportional to sensation magnitudes and that some other, nonlinear relationships could satisfy the results of cross-modal tests (MacKay, 1963; Ekman, 1964; Triesman, 1964; Zinnes, 1969; Shepard, 1978; cit. Gescheider, 1997). The logarithmic function proposed by Fechner attracted special attention. If sensations were related to physical stimuli by such a function, we could write in mathematical terms

$$\psi_1 = a \, \log\left(\frac{\phi_1}{\phi_{o1}}\right) \tag{1.14}$$

for one sense modality and

$$\psi_2 = b \, \log\left(\frac{\phi_2}{\phi_{o2}}\right) \tag{1.15}$$

for the other, where $\psi_{1,2}$ denote the sensation magnitudes and $\phi_{1,2}$, the physical-stimulus magnitudes referred to some standard magnitudes $\phi_{o1,2}$. By equating the sensation magnitudes, we would obtain

$$a \, \log\left(\frac{\phi_1}{\phi_{o1}}\right) = b \, \log\left(\frac{\phi_2}{\phi_{o2}}\right) \tag{1.16}$$

which can be rewritten in the form

$$\left(\frac{\phi_1}{\phi_{o1}}\right)^a = \left(\frac{\phi_2}{\phi_{o2}}\right)^b \tag{1.17}$$

or

$$\left(\frac{\phi_2}{\phi_{o2}}\right) = \left(\frac{\phi_1}{\phi_{o1}}\right)^{a/b} \tag{1.18}$$

Obviously, this is a power-function relationship. It shows that such a relationship can be obtained in the presence of logarithmic relationships between sensory stimuli and the sensation magnitudes they evoke. A power-function relationship in cross modal matching does not prove by itself that sensation magnitudes grow as power functions of adequate stimulus intensities. Evidence that they do, nevertheless, is introduced in a subsequent section.

Although cross-modal matching failed to prove conclusively the Power Law, it did provide two insights of enormous importance. As already mentioned, it demonstrated the generality of the concept of sensation magnitude and also implied an invariance of the functional relationship between numbers and this magnitude. Introduction of absolute scaling to cross-modal matching added evidence for intermodality constancy of subjective units. In terms of Eq. (1.13), not only the exponent ratio α/β but also the coefficient ratio k_1/k_2 became significant. A first test of mutual consistency, or transitivity of absolute scaling was performed in an intra-modal rather than a cross-modal experiment. The loudness of a 1 kHz tone was determined in the presence and absence of a masking noise by the numerical magnitude-balance procedure, as is described in Sect. 3 of this chapter (Hellman and Zwislocki 1964). According to the obtained results, masking altered both the shape and the position of the loudness function. Subsequent, direct loudness matches confirmed the magnitude-balance results to a close numerical approximation with respect to the intensities of the masked and unmasked tones required for loudness equality. The algebraic principle of transitivity was satisfied. According to this principle, when $a = c$ and $b = c$, $a = b$.

The intramodal transitivity tests were extended to the loudness relationships between binaurally and monaurally presented tones (Hellman and Zwislocki, 1963) and between tones of different frequencies (Hellman and Zwislocki, 1968). Further, they were extended to touch. In one experimental series, vibrotactile stimuli were presented to the thenar eminence of the right hand at 10 SLs by means of a cylindrical contactor of 2.9-cm^2 cross-sectional area protruding through a round opening in a rigid table surface (Verrillo et al., 1969). They consisted of bursts of sinusoids at 10 frequencies ranging from 25 to 700 Hz. The bursts were 600 msec long and were separated by 1,400-msec time intervals. The experiments included 6 observers and begun with absolute magnitude estimation and production whose results were geometrically averaged to obtain numerical magnitude-balance values for the group. Subsequently, every subject matched the subjective vibration magnitudes at all the frequencies to those produced at one of the two standard frequencies of 64 and 250 Hz, respectively, by the method of adjustment. The data obtained with the two standards were averaged. Finally, frequency contours of equal sensation magnitudes were constructed from the data of numerical magnitude balance and compared to those of heterofrequency magnitude matching. A sample of the results is shown in Fig. 1.23, where the solid lines connect the data points belonging to numerical magnitude balance, and the intermittent lines, those belonging to heterofrequency magnitude matching. Detection thresholds obtained in the two experimental series are also included. Clearly, the results of the magnitude balance agree approximately with those of direct magnitude matching, so that the requirement of transitivity is multiply satisfied within the experimental errors.

The transitivity of the results of absolute scaling was extended to intermodal relationships by means of cross-modal matching. In three series of experiments, the transitivity was demonstrated for vision, hearing, and touch. The results were corroborated by matching tactile pressure to loudness and subjective magnitude of vibration (Bolanowski et al., 1991).

In the first series, absolute magnitude estimation was applied to brightness and loudness. Subsequently, brightness and loudness magnitudes were matched directly to each other. Light stimuli consisted of 10-msec flashes of achromatic light produced by a Tungsten-filament bulb (750 W DDB, Sylvania, 100 V, color temperature $-2,800\,°K$) on a black background and subtended a solid angle of $1.8°$. Their intensity was controlled by selectable neutral-density filters and a neutral-density wedge that allowed continuous variation of the intensity. The light source was viewed binocularly, and the light flashes were presented at time intervals that depended on light intensity to minimize light adaptation. Before the beginning of an experimental session, the observers were dark-adapted for 30 min. Visual thresholds were

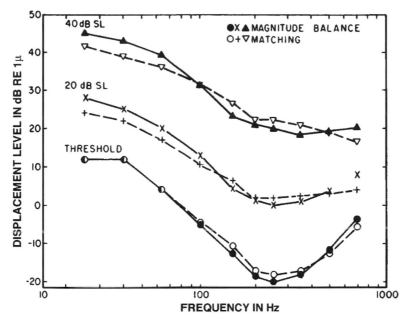

Fig. 1.23 Contours of constant sensation magnitudes of vibrotactile stimuli determined by direct inter-frequency matching (intermittent lines) and derived from numerical magnitude balance at each frequency (solid lines). The bottom curves indicate the respective thresholds of detectability. The threshold at 250 Hz served as reference for the indicated SLs. A group of 6 observers participated in the experiments. Reproduced from Verrillo et al. (1969) with permission from the Psychonomic Society

determined by the method of limits and served as reference. All the data of this and following experiments were recorded as functions of individual SLs to minimize intersubject variability (Hellman and Zwislocki, 1961).

In the auditory part of the experiment, the stimuli consisted of 20-msec bursts of a 1,000-Hz tone with 10-msec rise- and fall times that made the associated transients inaudible. The stimuli were presented to the right ear by TDH 39 earphones mounted in MX-41 cushions, and their intensity was varied in 2-dB steps. The threshold of audibility was determined by the method of limits with the tone bursts repeated at a rate of one per second. As for vision, the stimuli were presented singly for magnitude estimation and cross-modal matching. The time-consuming procedure was followed to avoid light adaptation and, at the same time, achieve intermodal symmetry.

A group of 5 initially inexperienced observers participated in the experiments. They were required to make absolute estimations of both loudness and brightness magnitudes in three sessions comprising two blocks of trials each. The loudness-estimation blocks of trials preceded the brightness-estimation blocks to utilize part of the time required for dark adaptation. The first two sessions served as training sessions. Geometric means of the individual data of the two blocks of the third session were calculated for both vision and hearing and used to obtain geometric

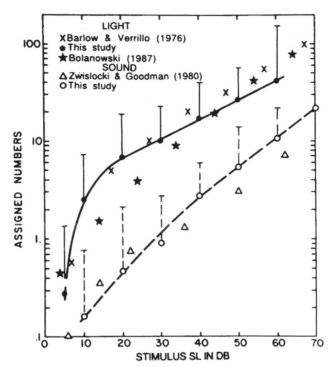

Fig. 1.24 AME of brightness (*filled circles*) (achromatic light, 1.8° solid angle, 10-msec flashes) and loudness (*unfilled circles*) (1,000 Hz, 20-msec bursts). Geometric means of a group of 5 observers. The curves have been fitted to the data points by eye; the vertical lines indicate the population standard deviations. The crosses indicate AME values of brightness for an achromatic ganzfeld illumination, and the asterisks, the values obtained for achromatic light flashes of 1-s duration and 2° subtended angle. The triangles indicate the AME loudness values of Fig. 1.19 (*filled circles*). Reproduced from Bolanowski et al., (1991) with permission from Taylor & Francis Group, LLC

means and standard deviations for the group. The latter are plotted in Fig. 1.24 by filled circles for vision and unfilled circles for hearing. The vertical bars above the circles indicate the intersubject standard deviations. The smooth curves approximating the data points have been drawn by hand. Interestingly, the same SL produced substantially greater numerical estimates of brightness than of loudness. The difference is associated with a much faster growth of brightness than of loudness near the respective detection thresholds. At higher SLs, both data sets follow approximately power functions that appear as straight lines on the double-logarithmic coordinates of the figure. At these SLs, the brightness curve is somewhat flatter than the loudness curves, indicating a somewhat smaller power exponent. Both exponents are probably somewhat biased because they have resulted from AME rather than numerical magnitude balance, but the bias should be similar for both and should practically disappear in their mutual relationship.

Because brightness and loudness estimates occurred in the same sessions, their mutual interaction became of concern. For this reason, the obtained results were compared to relevant results of prior experiments in which brightness and loudness were scaled separately by the AME method. Unfortunately, no experiments with the same stimulus parameters were found for vision. The included results of Barlow and Verrillo (1975) (crosses) were obtained on 6 observers with Ganzfeld light flashes of 10-msec duration. According to Mansfield (1973), the target size beyond 1° of solid subtended angle has little effect on the shape of the brightness function. This is confirmed by the approximate agreement of their data with those of Bolanowski et al. Although their data suggest a somewhat steeper curve slope at high SLs, the difference is certainly within the variability encountered among magnitude-estimation experiments. The loudness results were compared to those of Zwislocki and Goodman (1980) obtained on 12 observers with the same stimulus parameters, except that the tone bursts were repeated at a rate of one per second. Their data are indicated by the triangles, which show approximate agreement with the data of Bolanowski et al. indicated by the unfilled circles. Thus, neither the subjective brightness units nor those of loudness appear to have been affected appreciably by scaling both modalities in the same session.

At the conclusion of the magnitude-estimation sessions, the observers were requested in the Bolanowski et al. study to match the brightness and loudness magnitudes in a separate session. To minimize the effect of visual adaptation, the following method was introduced. The light and sound stimuli were presented in separate pairs, and the observers had to judge which stimulus appeared to have the greater magnitude according to a forced-choice procedure. Using the individual magnitude-estimation curves, a tone SL was selected arbitrarily and a vertical line was drawn at its coordinate to meet the loudness curve, as shown in Fig. 1.25 by the dashed lines for several SLs. From the point of intersection, a horizontal line was drawn to intersect with the brightness curve. The point of intersection determined the initial light intensity given in the forced-choice procedure. If this intensity produced the response "light stronger" (L.S.), the light intensity was decreased by a small step for the next stimulus-pair presentation. It continued to be decreased until the observer responded "tone stronger" (T.S.). Thereupon, the light intensity was increased, and so forth. The middle of the intensity range between the response reversals was accepted as the light intensity at which the magnitudes of brightness and loudness appeared to be equal. If the first response was "T.S.," the light intensity was increased in steps until the response "L.S" occurred, and so forth. The result of the procedure for a typical observer are shown in Fig. 1.25 together with the corresponding results of AME. The ranges between the "L.S" and "T.S." responses are indicated by arrowheads and short vertical lines. As can be seen in the figure, they agree approximately with the AME of brightness. The forced-choice procedure was repeated for all the observers, and its results are shown in Fig. 1.26 by a different symbol for every observer. They are compared to the theoretical cross-modal matches constructed from the magnitude-estimation curves of Fig. 1.24 averaged over the group of the observers. The predicted SLs for light and tone expected to produce cross-modal magnitude equality are indicated by stars interpolated by the solid

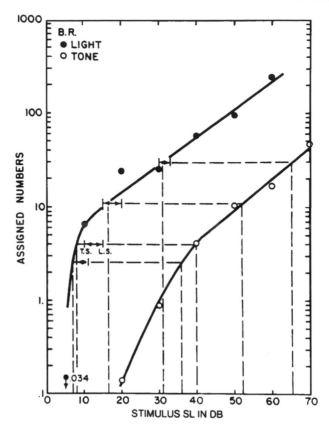

Fig. 1.25 Individual data of AME of brightness and loudness obtained under the conditions of Fig. 1.24. Dashed lines indicate SLs of light and sound expected to produce equal subjective magnitudes of brightness and loudness. Arrows indicate ranges of light intensity within which the brightness magnitude could not be clearly distinguished from loudness magnitude. Reproduced from Bolanowski et al. (1991) with permission from Taylor & Francis Group, LLC

line. Clearly, the data points indicating empirical individual magnitude matches are scattered close to the theoretical curve, indicating that the observers used approximately the same subjective units in AME for both brightness and loudness. The agreement also validates the course of the brightness curve relative to that of the loudness curve and, in so doing, validates the finding that, under the experimental conditions of Bolanowski et al., the brightness magnitude was greater than the loudness magnitude by about one order of magnitude in the presence of equal SLs. This somewhat surprising finding could not have been made by ratio scaling with arbitrary units. It is true that a closer scrutiny of the data of Fig. 1.26 suggests that, in AME, brightness tended to be slightly overestimated relative to loudness in mid SL range but the bias is equivalent to only about 3 dB in SL terms.

A second series of Bolanowski et al. experiments concerned loudness and the subjective magnitude of vibration. The experiments were somewhat easier to

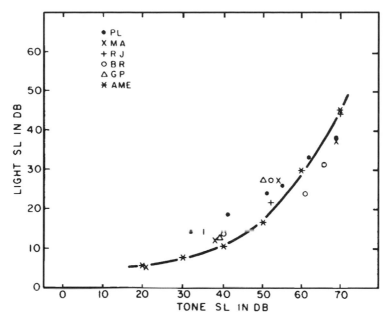

Fig. 1.26 SLs of light and sound producing equal subjective magnitudes of brightness and loudness, as derived from the AME group data of Fig. 1.24 (*asterisks and solid line fitted to them by eye*) and as obtained by individual observers through intermodal magnitude matching. Reproduced from Bolanowski et al. (1991) with permission from Taylor & Francis Group, LLC

perform than in the first series, because they were not encumbered by adaptation, so prominent in vision experiments. The stimuli for loudness were similar to those used in the first series, except that they were repeated at a rate of one per second. The timing of the tactile stimuli was the same. The auditory stimuli were delivered through an insert phone to the right ear. A similar dummy earphone was placed in the contralateral ear to exclude the sound produced by the tactile vibrator (Zwislocki et al., 1967). The vibration stimuli were delivered according to the method of Verrillo (1963) to the distal pad of the left middle finger through a cylindrical contactor having a cross-sectional area of 0.29 cm². The contactor protruded through a round opening in a rigid surface leaving a 1-mm gap surrounding the contactor and indenting the skin by 0.5 mm. A group of 8 initially inexperienced observers participated in the experiments, and the psychophysical methods were essentially the same as in the first series of Bolanowski et al. However, because of negligible adaptation, blocks of trials alternated between sound and vibration in every testing session. Also, to counterbalance the sequence of the first experimental series, the matching experiments preceded the AME experiments. In the former, the auditory and tactile stimuli alternated while an observer performed the matching procedure by adjusting the intensity of the first or the second by means of a round, unmarked knob. The stimulus under the observer's control was alternated among the sessions. The group results of AME are shown in Fig. 1.27 by filled circles

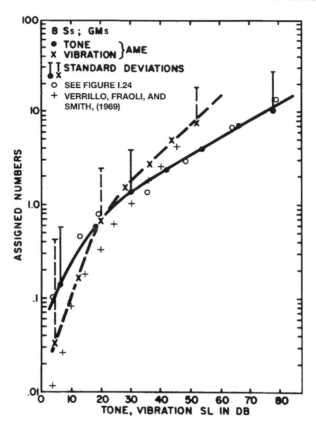

Fig. 1.27 AME of loudness (20-msec bursts, 1,000 Hz) (*filled circles*) and tactile vibration (20-msec bursts, 120 Hz) (*slanted crosses*) obtained on a group of 8 observers. The curves were fitted by eye to the data points, and the vertical lines indicate the standard deviations. For comparison, unfilled circles show the AME data of loudness indicated in Fig. 1.24 by triangles, and vertical crosses, numerical magnitude-balance data for 150-Hz vibration (Verrillo et al., 1969). Reproduced from Bolanowski et al. (1991) with permission from Taylor & Francis Group, LLC

for loudness and by slanted crosses for the subjective magnitude of vibration. The solid and intermittent lines interpolate approximately the respective data points. To demonstrate that the cross-modal magnitude matching did not affect the AME results appreciably, they are compared to similar results obtained in past experiments dedicated to the one or the other modality but not to both, and that did not include cross-modal matching. The auditory data points indicated by the unfilled circles are the same as in Fig. 1.24 (Zwislocki and Goodman, 1980); the tactile ones, indicated by straight crosses, have been derived from the study of Verrillo et al. (1969). They were obtained at a vibration frequency of 150 Hz instead of 120 Hz by the method of numerical magnitude balance rather than AME alone. There is practically no difference between the vibro-tactile magnitude functions at 150 and 120 Hz. However, numerical magnitude balance tends to produce somewhat lower

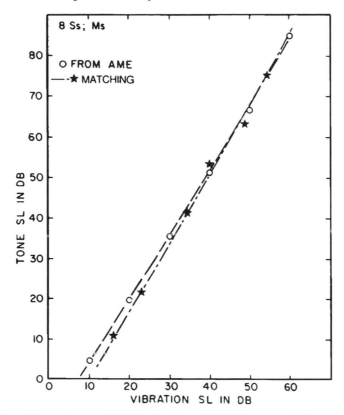

Fig. 1.28 SLs of 20-msec bursts of 120-Hz vibration and 1,000-Hz tone producing equal subjective magnitudes, as derived from the AME data of Fig. 1.27 (*unfilled circles*) and as obtained by direct magnitude matching (*stars*). Reproduced from Bolanowski et al. (1991) with permission from Taylor & Francis Group, LLC

magnitude values than AME. This difference is evident in Fig. 1.27. Otherwise, the good agreement between the corresponding results of the Bolanowski et al. and the two preceding studies suggests that cross-modal matching preceding the AME experiments did not affect appreciably the results of the latter. Also significant is the agreement between the loudness functions of the first and second of the Bolanowski et al. series, as indicated by the agreement of both with the results of Zwislocki and Goodman.

A comparison of cross-modal matches derived from the Bolanowski et al. data of Fig. 1.27 to corresponding direct matches is shown in Fig. 1.28. The former are indicated by the unfilled circles and the intermittent line, the latter by the stars and the dash-dot line. In view of the known intra-observer and inter-observer variability, the difference between the two sets of data must be considered as negligible. Thus, intermodal transitivity of AME for loudness and the subjective magnitude of vibration holds at least for group averages.

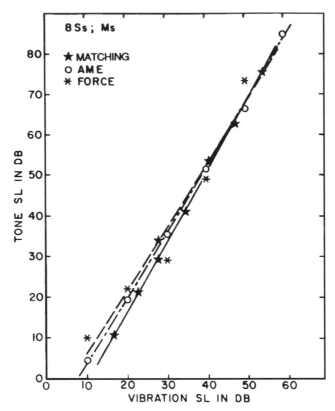

Fig. 1.29 SLs of the same stimuli as in Fig. 1.28 that produced equal sensation magnitudes, as determined by matching to subjective active finger pressure (asterisks and intermittent regression line), compared to the results of Fig. 1.28 (*circles and stars*). Reproduced from Bolanowski et al., (1991) with permission from Taylor & Francis Group, LLC

The third series of Bolanowski et al. experiments differed from the second only in that AME was replaced by pressure exerted by the right index finger on a knob protruding through an opening in a table surface. The resulting physical force was measured with a calibrated electrostatic device. The experiments were performed on 8 observers. The group averages of the tone and vibration SLs producing equal sensation magnitudes, as derived from equal finger pressures, are shown in Fig. 1.29 by asterisks. They are compared to corresponding data resulting from direct matches and from the AME experiments. The latter are reproduced from Fig. 1.28. Clearly, all the data of Fig. 1.29 are in approximate agreement with each other, demonstrating mutually consistent transitivity for both AME that involves numbers, and pressure matching that does not. Because group data can hide mutually canceling characteristic features of individual data, a typical example of the latter is shown in Fig. 1.30. The symbols are the same as in Fig. 1.29. The agreement among the three sets of the data points confirms the group results of that figure. Note that the curve approximating the results of direct cross-modal matches asymptotes on a power

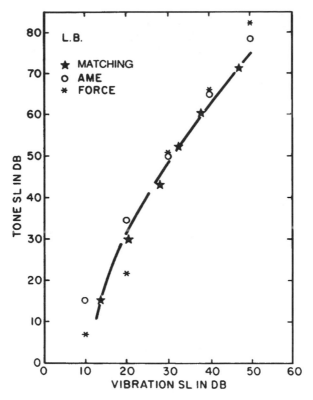

Fig. 1.30 Results obtained on one observer for the same experimental conditions as in Fig. 1.29. Reproduced from Bolanowski et al. (1991) with permission from Taylor & Francis Group, LLC

function at high SLs of vibration, and that this is also true for the data derived from AME and pressure matching.

A study especially dedicated to individual behavior in cross-modal matching derived from AME was performed by Collins and Gescheider (1989). Their psychophysical procedure was similar to that of Bolanowski et al. As stimuli they used 1,000-Hz tone bursts of 1,250-msec duration repeated every 2.5 s and horizontal lines projected on a screen by an overhead projector at a distance of 2.5 m. The lines were varied continuously in length between 0.25 and 125 cm by a variable-aperture device mounted on the projector and under either the experimenter's or the observer's control. The experiments were performed on 12 adults and 12 children 4–7 years of age. One child, 4 years old, was excluded from the data analysis because she was unable to order the numbers properly, when tested. The order of procedures was counterbalanced in that half of the observers first matched the line length to loudness, then, performed AME of line length followed by AME of loudness. For the other half the order was reversed. Power functions were fitted to all the AME data according to the least-squares method, then, compared to the results of direct cross-modal matching. For methological details see the article of Collins and Gescheider (1989). The results for the adults are displayed in Fig. 1.31,

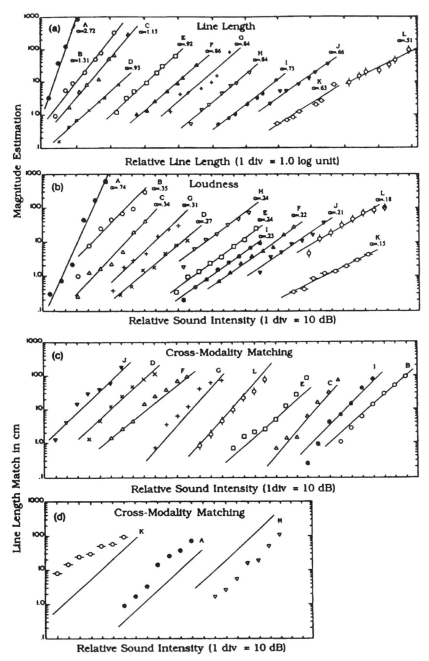

Fig. 1.31 AME of line length and loudness of 1.250-s 1,000-Hz tone bursts, and crossmodal matching between their subjective magnitudes. In the upper two panels, the straight lines indicate the power-function least-square approximations of the individual data obtained on 12 adult observers. In the lower two panels, the data points indicate direct matches, the straight lines, matches derived from AME. The lowest panel shows data of 3 observers who did not maintain the same subjective units for AME of line-length and loudness. Reproduced from Collins and Gescheider (1989) with permission from the American Institute of Physics

the ones for AME in the upper two panels, those for cross-modal matches, in the lower two. In the latter, the solid lines indicate the matches predicted from AME, the data points, the matches obtain directly. Both sets are consistent with each other for 9 out of 12 observers. For three observers, the predicted and empirical matches disagree in the way suggesting that the observers used different subjective units for the line-length and loudness magnitudes in AME. Similar results are displayed in Fig. 1.32 for the children. The predicted and empirical cross-modal matches are in approximate agreement for 9 out of 11 children. For two, they disagree in a way similar to that of the three adults.

Overall, the results of the Bolanowski et al. and of Collins and Gescheider experiments indicate that AME as well as AMP are consistent with cross-modal matching and, in this way, pass the test of transitivity in almost all observers. An additional confirmation of transitivity on an individual basis was obtained by Hellman and Meiselman (1988, 1990, 1993) who extended it to observers with hearing loss and, therefore, altered loudness functions. The next section provides the foundation for the human ability to make consistent and accurate cross-modal matches and demonstrates that such matches, as well as numerical estimates of AME and AMP, constitute measurement.

1.5 Relevant Theory of Measurement

To establish an empirical function, like the power function, both the values of its independent and dependent variables must be measured. In Stevens' power-law functions, the former are determined by means of well established physical methods of measurement. The latter, on the other hand, are determined by means of the newly introduced method of magnitude estimation or its variants, in particular, the absolute magnitude estimation. Do these methods constitute legitimate measurement? The question can only be answered if we agree upon an unambiguous definition of measurement. In the past, measurement was defined by a blue-ribbon committee of scientists and philosophers as "assignment of numbers to things or events according to rules" (cit. Stevens, 1951). Different rules were suggested by different authorities, so that this definition cannot be regarded as unambiguous. In addition, it misses the fundamental empirical operation of measurement, which consists of matching common attributes of things or events (Zwislocki, 1991). Already Stevens emphasized in his "Psychophysics" (1975) that matching is involved in every measurement. Here, I attempt to show that useful measurement can be performed without explicit use of numbers, in particular, numbers denoted symbolically by numerals. The latter are only necessary for counting the units of measurement, when such units are defined, and for mathematical modeling of the empirical world. For example, Stevens' power low cannot be expressed without their use.

In a past article, I gave examples of measurement without numbers (Zwislocki, 1991); here I add other anecdotal examples taken from every-day life. The first concerns joyous activities of Christmas time. My wife, Marie, and I were returning

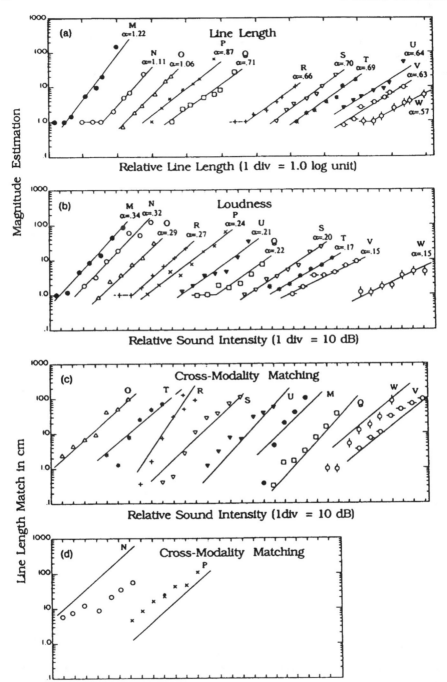

Fig. 1.32 The same experiment as in Fig. 1.31 but performed on 11 children. The results for two children shown in the lowest panel indicate a disagreement between direct cross-modal matches and matches derived from AME. Reproduced from Collins and Gescheider (1989), with permission from the American Institute of Physics

from Christmas shopping when we noticed on the right side of the street a small stand of Christmas trees. Although we did not plan on it previously, we decided to take a look and see if a tree suitable for us was hiding among the many that were exhibited. After a few minutes of intense searching, we found one that could possibly satisfy our needs. We wanted a shapely tree that would almost, but not quite, reach the ceiling of our living room. The particular tree seemed to be just tall enough, but we were not sure. We were not sure how exactly we remembered the height of the room nor did we have a measuring tape to measure the height of the tree. Fortunately, the salesman had a ball of string and let us cut a piece of it equal to the tree height. He agreed to reserve the tree for us for a half an hour. We rushed home and compared the length of the string to the height of the living room. Our intuition was right – the string was just a little shorter than the room height. We quickly returned to the tree stand to pick up the tree that, fortunately, was still there.

By matching the string length to the tree height, then comparing it to the room height, did we not make two measurements, useful measurements, without any use of numbers? One could claim that the number was 1, but this would have no meaning since the tree and, therefore, the string could have been of any length. I think that measurements of this kind abound in life. Does a tailor use a measuring tape or any numbers to adjust the length of trousers you are buying to the length of your legs? When cooking, don't you add a dash of this or a dab of that to make whatever you are cooking taste just right? Isn't this a kind of measurement? You could achieve the same result by weighing the quantity used with a calibrated scale or measure it with a calibrated cup. An analogous situation arose when my wife and I compared "by eye" the height of the Christmas tree to the height of our living room, then verified our subjective impression with the help of a string.

There are measurements of an apparently different kind that occur without numbers. A thermostat, especially an uncalibrated one, provides a typical example. When attached to an electrical space heater, such a thermostat turns on the current every time the temperature in a room sinks to a certain level and turns it off whenever the temperature increases to a certain higher level. To perform these operations, the thermostat must be able to measure the temperature. The measurement is internal and its mechanism unknown to a naive observer. Nevertheless, he or she is able to set the thermostat to a comfortable temperature. Such temperature is also determined by an internal mechanism – this one within ourselves. Its mechanism is not yet entirely known, but one thing is certain: to perform its function, it must be able to measure the temperature, just like in the instrumental thermostat – our organism must be able to function as a measuring device.

More generally, the ability to make measurements must be an inherent property of living organisms. Its effects can be observed throughout the animal kingdom. How could a squirrel jump successfully from one tree branch to another if it were not able to measure the distance between the branches and match its muscular effort to it? The same goes for monkeys and birds. An owl is able to catch a mouse, its preferred pray, in darkness by determining the coordinates of its location with the help of directional hearing. Humans have to perform similar measurements. We measure subjectively the width of a ditch before we decide to jump over it, we have to determine the location of a target at which we want to throw a ball or a spear. Reliable measurements are performed even by lower organisms. In an earlier publication (Zwislocki, 1991), I cited an article of Schmidt and Smith (1987) in which they describe parasitoid wasps (*Trichogramma*) that are able to measure the size of insect eggs in which they deposit their own. The wasps are minute, ranging in size from 0.3 to 0.8 mm, and use host eggs 0.3–2.5 mm in diameter. The authors found that the wasps lay a constant number of eggs per unit volume of a host egg, apparently, to maximize the survival of their larvae. Since the number of eggs does not depend on the wasp size, a measurement of the size of a host egg relative to the wasp size is excluded. Instead, the wasps explore the potential host egg by walking over it and touching it with their antennae. According to the authors, only one parameter of the wasps' behavior is directly proportional to the number of eggs laid. It is the time needed by a wasp to walk over a host egg. The time does not depend on the wasp size. The authors suggest that the wasps compare this time to the time set by an internal clock. Of course, biological clocks of various kinds are well known.

The above examples should demonstrate sufficiently that animals use subjective measurement to cope with their environment. I called such a measurement "natural measurement" (Zwislocki, 1991). Of course the measurement cannot be direct but must operate on internal representations of the external variables – ergo, the representations must be measurable. Visual distance is such a representation, the subjective impression of size is another; but so is the feeling of muscular tension and the brightness of light and the loudness of sound, which are part of our world of sensations. One of the most fundamental features of the natural internal measurement is the ability of living organisms to compare the magnitude of one variable to the magnitude of another variable that is qualitatively entirely different. Comparison between the impression of the distance to an object and the sensed muscular effort necessary to reach the object is one example. Comparison between the conflicting feelings of hunger and fear can often be inferred from animal behavior. In psychophysics, cross-modal matching introduced by Stevens (1959) as a systematic measurement is an example of particular interest here (see Stevens, 1975, and Gescheider, 1997, for reviews).

In physical measurement that may be regarded as the model of all measurement, only common attributes of things or events can be matched – length can only be matched to length, weight to weight, time interval to time interval. How then, in psychophysical experiments, brightness can be matched to loudness or the subjective intensity of vibration to loudness? To the best of my understanding, there is only one possibility. Humans, and animals also, must be able to abstract from sensations having different qualitative attributes an abstract sense of magnitude common to them all. In cross modal matching, this abstract sense of magnitude must be what is compared. As a consequence, we should not instruct observers to match loudness to brightness, for example, but rather, to match the magnitude of loudness to that of brightness. More generally, we should instruct them to match the magnitude of one sensation to the magnitude of another sensation rather than one sensation to another (Zwislocki and Goodman, 1980). These statements are in agreement with the definition of measurement as an operation in which common attributes of things or events are matched. Stated as an explicit definition, "measurement consists of matching common attributes of things or events." Restriction to "common attributes" is essential and distinguishes measurement from other matching operations that can have innumerable aspects. For example, matching the stile of chairs to the stile of a table or matching the architecture of a house to the surrounding countryside would not be regarded as measurement.

How do numbers enter into measurement? Actually, the process is very simple. First of all, numbers come in to express numerosity of things or events that occur naturally – we can count sheep, horses, people, or houses; we can count how many car accidents take place every day, how many days there are in a year, etc. They can also be used in connection with continua as soon as we define a unit of measurement. Then, we ask how many units are contained in the entity we measure. We express the length of a table in terms of the number of units of length, such as inches or centimeters. We express the weight of a person in terms of the units of weight, such as pounds or kilograms. According to convenience, we can define large units or small units, meters, or millimeters, for example. Because of this arbitrariness, numbers, as symbolized by numerals, have no absolute meaning in measurement. As a matter of fact, they are meaningless unless we define the underlying unit – 100 cm is no longer than one meter.

Although numbers enter the process of measurement in a simple way, their significance is enormous because the whole of mathematics used to represent quantitatively the world around us and within us is based on them. No quantitative modeling can take place without them. Psychology and psychophysics are no exceptions.

Numbers are required to define Stevens' power law. For this purpose, both the physical and the corresponding psychological variables must be expressed in their terms. The physical variables can be determined quantitatively by well-established methods of numerical measurement, using established units, but how can numbers be coupled to psychological continua? How is it possible to define a psychological unit of measurement? As mentioned above, Stevens (1956) thought at first that a unit of sensation magnitude can be defined arbitrarily by producing a sensation magnitude with an appropriate stimulus and assigning a number to it. However, he soon found that the measured relationship between the sensation magnitudes and stimulus intensities that produced them depended nonlinearly on the units so defined. He then, requested his observers to implicitly define their own units by assigning a number to the sensation magnitude produced by the first stimulus they received. They were to refer to these units throughout the rest of the experiment. Going one step further on the basis of our own experiments, Hellman and I came to the surprising conclusion that numbers acquire absolute psychological magnitudes that are the reason for response biases resulting from the imposition of reference standards conflicting with these magnitudes (Hellman and Zwislocki, 1961). The conclusion, which implies a natural unit, has been confirmed on many occasions. Some examples have been given above (e.g. Hellman and Zwislocki, 1963, 1964; Zwislocki and Goodman, 1980).

Whereas numerous experiments have shown that, in magnitude estimation and production, observers tend to use absolute numerical scales, demonstration how such scales are formed is another matter. The experiments on children discussed above suggest that the scales are formed early in life, probably, when children learn numbers by counting familiar objects (Zwislocki and Goodman, 1980). Such counting leads to a numerosity scale that is absolute.

To apply the numerosity scale to continua, a transition is required that may be influenced by the size of the objects on which children learn the meaning of numbers. If they count relatively large objects, the sum total of the objects may appear relatively large and relatively small numbers are applied to continua whose scaled portions may appear relatively small. On the other hand, if they count relatively small objects, their sum total may appear relatively small and the scaled portions of continua relatively large by comparison. Under such conditions, relatively large numbers would be expected to be used. On several occasions, I asked people belonging to various walks of life which number they thought was of medium size, not large or small but medium. The answers ranged from 3 to 12 with a median of 7. The people did not think that my question was silly. This means that they had developed absolute psychological scales of numbers in spite of frequent use of relative number scales in counting money or measuring lengths. The inter-individual differences in the subjective magnitudes of numbers can explain partially the inter-observer variability in magnitude estimation and production experiments.

A fundamental aspect of the transition from numerosity to a psychological continuum is that psychological magnitudes of numbers are probabilistic rather than deterministic. In other words, the psychological magnitude of a number is not entirely fixed but is statistically distributed, just like other psychological magnitudes. Such distributions are clearly evident in the variability of the observers' responses. They suggest that psychological distributions of the magnitudes of successive cardinal numbers tend to overlap, so that, often, two different numbers are assigned to a psychological magnitude produced by the same physical stimulus. The statistical diffusion of the psychological magnitudes of numbers converts the digital nature of numerosity to a quasi-continuum. The digital nature of numerosity and its probabilistic transformation to a psychological continuum may explain another aspect of response variability in magnitude estimation and production experiments. Consistently, the variability is greater at low stimulus levels eliciting numerical responses in terms of fractions than, at higher levels, eliciting whole number responses. Children learn fractions as a separate concept, after they had learned whole numbers. Subsequently, in everyday life, fractions are used less often than whole numbers. Furthermore, association of fractions with a continuum that is in reality infinite is paradoxical and may evoke a good deal of psychological uncertainty in spite of the certainty that a fraction on an absolute scale is smaller than unity. It is often seen that naive observers, performing magnitude estimations for the first time, use relatively large numbers in an intuitive effort to avoid fractions. Once they are induced to use fractions, for example, in association with very short lines, their numerical estimates decrease and fall in line with those of experienced observers. They also converge on the results of magnitude production in which the numbers are dictated by the experimenter (e.g. Zwislocki and Goodman 1980). In addition to being required by observers for the expression of small psychological magnitudes, fractions are instrumental in making the essentially discrete distribution of numbers approach a smooth continuum. Observers often use numbers such as one and one quarter or two and a half, and so forth.

Summarizing, I feel that the following conclusions are warranted: Measurement consists of matching common features of things or events and, except for numerosity, can be performed without the use of numbers.

When numbers are introduced, they enter psychological measurement in a different way than they enter physical measurement – in the latter, they form ratio scales with arbitrarily defined units; in the former, predominantly, absolute scales based on subjective magnitudes of numbers that may differ from observer to observer.

1.6 Validity of the Power Law

According to the power law, sensations and perhaps other psychological magnitudes follow power functions of stimulus magnitudes that evoke them. When subjective magnitudes of numbers are matched to sensation magnitudes, such as those of loudness or brightness, the numbers tend to obey power functions of the stimulus magnitudes associated with the sensations. Does this mean that the sensations themselves are power functions of stimulus intensity? Not necessarily, according to Atteneave (1962) and some others (e.g. McKay, 1963; Treisman, 1964) who suggested that subjective magnitudes of numbers may be nonlinearily related to the numbers. In the section on cross-modal matching, a simple mathematical derivation showed that empirically obtained power-function relationships among stimulus variables producing equal subjective magnitudes of different modalities do not prove that these magnitudes are related by power functions to the stimulus variables. In view of the resulting indeterminacy, Stevens (1975) maintained that the latter relationships are not essential as long as the hetero-modality stimuli are related to each other by power functions. Such a concept leaves the internal world of sensations and, possibly, other psychological attributes with an unknown relationship to the outside world. In addition, it contains an inconsistency because the inter-modal relationships are expressed in terms of stimulus variables rather than sensation variables.

Fortunately for Stevens' power law, the suggestion that subjective impressions of numbers may be nonlinearly related to numbers contradicts the fact that numbers, as denoted by numerals, have been defined so as to directly express magnitude relationships. We are taught to assign a numeral "two" to something that is twice as long as a defined unit length, a numeral "three" to something that is three times as long, etc. We will say that a table is twice as long as another table, whether the lengths have been measured with a ruler or by eye. There is no apparent reason why we should use the number system in a different way in reference to the objective physical world and to its reflection in our subjective impressions. When a sound appears to us five times as loud as another sound, we will say so, the same for the brightness of light or the strength of an odor. The similarity between the applications of our numerical system to the "objective" physical world and the internal "subjective" world is nicely demonstrated by the following mental experiment. Let us project a short line on a screen, then another line of the same physical length out of alignment with the first. Then, let us align the second line with the first and place both end-to-end. The resulting line, as measured by a ruler will be twice the length of each component line. Of course, the principle of additivity holds. Now, let us cancel the second line, then, reintroduce it again but, this time, make its length the same as of the first line by eye. The physical equality is likely to be less accurate than in the first instance, but experience with AME and AMP of line length tells us that the error will not be great. The decreased accuracy will be due to the fact that, instead of matching the physical lines, their subjective impressions will be matched. The two component lines can be aligned and placed end-to-end again entirely under visual control. Their lengths will be added to each other and the resulting length will appear approximately twice as great as of the component lines. We will express this impression by assigning to the total length a number approximately twice as great as assigned to each component line. We can verify this result by an auxiliary experiment with AME and AMP of subjective line length as a function of the objective line length or refer to the results of one of several past experiments (Verrillo and Irvin, 1979; Zwislocki and Goodman, 1980; Verrillo, 1981). A particularly striking result can be obtained by summing lines of subjective unit length. On the average, observers assign number one in AME or AMP experiments to physical line length of about 4 cm. Now, when two line lengths of subjective unit length are added, a line length of two subjective units is obtained and a number of ~ 2 is assigned to it. When a third line of subjective unit length is added, the number of ~ 3 is assigned to the total, and so forth. The numbers coincide approximately with the number of units, just like in physical measurement. Can there be any doubt that, at least on the average, numbers are assigned in AME and AMP in direct proportion to subjective magnitudes?

As the following examples demonstrate, the proportionality is not exact, however, and can be violated noticeably by individual observers. The examples show in the first place that approximate direct proportionality between subjective magnitudes and the numbers assigned to them is not limited to line length. In the second place, it shows that the proportionality is approximately true only on the average. Individual observers often couple numbers with subjective magnitudes according to power functions with exponents smaller or larger than unity, as judged from the numerals they choose. The departures from the unit exponent tend to be small, however (Zwislocki, 1983).

The examples concern loudness summation. That loudness of two sound stimuli processed independently sums linearly has been assumed since the 1930s on the basis of informal impressions (Fletcher and Munson, 1933, Fletcher, 1940). Independent processing appears to occur when one stimulus is presented to one ear and the other to the contralateral ear or the stimuli differ sufficiently in sound frequency. The summation phenomenon was used to construct loudness functions of sound intensity (Fletcher and Munson, 1933). The gross similarity of these functions with the functions obtained by ratio-scaling procedures indicates that the assumption of linear summation was consistent with the assumption that numbers are assigned in direct proportion to subjective magnitudes.

In more recent times loudness summation was studied explicitly in several experimental series (e.g. Levelt et al., 1972; Marks, 1978; Zwislocki, 1983) in part with the help of the non-metric conjoint measurement of Luce and Tukey (1964). In the example described below, pairs of short, 20-msec tone bursts separated by time intervals of 50 msec were presented monaurally, the first burst at a sound frequency of 4 kHz, the second at that of 1 kHz (Zwislocki, 1983). The frequency difference was sufficient to assure independent processing. The loudness of the two-tone bursts presented singly and in pairs was determined by AME. Prior to the auditory experiments, the 5 observers who successfully completed them had to scale the subjective length of lines projected on a screen in random order of physical length. Since the observers never participated in scaling experiments before, the line-length experiment allowed them to become acquainted with the AME procedure. Its results served subsequently as a reference for evaluating the relationships between subjective magnitudes and the numbers assigned to them. They are shown in Fig. 1.33 by the filled circles and the solid straight line fitted to them by the method of least squares ($r = 0.999$). The line follows a power function with an exponent $\alpha_L = 0.97$, almost unity. The crosses indicate the results of another experiment in which the same set of lines was used for informal line-length scaling at a small social gathering of 14 persons. The agreement between the two experiments documents the stability of the method. The results are also in agreement with those of Teghtsoonian and Beckwith (1976) and of Verrillo (1981). Note that number 1 in the subjective estimates coincides with a physical line length of about 4 cm. The same is approximately true for the line-length functions of Fig. 1.14 determined on adults and on children who did not know numbers below 1 and above 99.

Loudness magnitudes of the 1- and 4-kHz tone bursts were scaled separately in the next two sessions. The group results are shown in Fig. 1.34 and compared to the results obtained previously for 1-kHz tone bursts of the same duration. The scaling of the latter occurred by both AME and AMP (Zwislocki and Goodman, 1980). There is excellent agreement between the two studies, and the 1-kHz and 4-kHz functions practically coincide. There is also practically no difference between the AME and AMP data. This is expected of highly experienced observers, as suggested by Fig. 1.19. Such an agreement can also be found in groups of observers who scaled line length before scaling loudness.

In the experiments with burst pairs, the loudness ratios between the bursts of every pair were determined from individual data entering into the group data of Fig. 1.34. The following nominal ratios were chosen: $n_4/n_1 = 0.5$; 1; 1.5, where n_1 refers to the 1-kHz tone and n_4, to the 4 kHz one. As a control, the loudness of the 1-kHz bursts presented separately was scaled again in the sessions with burst pairs. In a counterbalanced procedure, the separate 1-kHz bursts were presented first in half the experimental runs and second, in the other half, in a randomized order. As is evident in Fig. 1.35, the loudness functions generated under all the conditions had similar slopes that corresponded to power-function exponents ranging

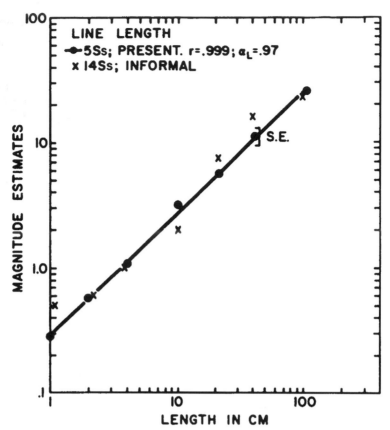

Fig. 1.33 AME of line length. Filled circles indicate geometric means of estimates made by 5 observers in a laboratory experiment, the straight line, their least-squares approximation, and the bracket, their standard error. The crosses indicate analogous results obtained informally on 14 observers at a social gathering. Reproduced from Zwislocki (1983) with permission from the Psychonomic Society

from $\alpha = 0.34$ to 0.39, with a mean of $\alpha = 0.37$. The mutual similarity of the exponents allowed all the data to be approximated by straight lines obeying the mean exponent. Because this exponent was somewhat larger than for the 1- and 4-kHz tone bursts presented separately in the preceding session, the nominal ratios n_4/n_1 proved not to be exact. Nevertheless, they were approximated reasonably closely by the mean experimental data producing the ratios of 0.6, 1.1 and 1.7, equivalent to the ratios $(n_4 + n_1)/n_1$ of 1.6, 2.1 and 2.7 indicated in the figure. On the average, the latter deviated by less than 1% from the nominal ratios.

The experiments on loudness summation and line-length scaling, made it possible to compute the individual and mean relationships between the subjective magnitudes and the numbers associated with them (Zwislocki, 1983). In the computations, linear additivity of loudness magnitudes was assumed, as verified by the group data of Fig. 1.35 and the nonmetric procedure of Luce and Tukey (1964). The

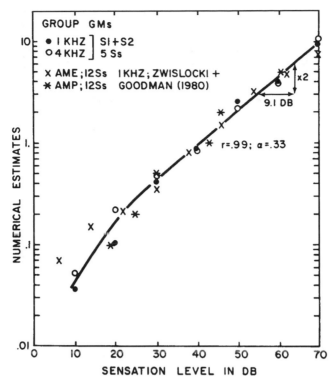

Fig. 1.34 AME of 20-msec tone bursts at 1,000 and 4,000 Hz produced by the same 5 observers as in Fig. 1.33. The circles indicate geometric means of two sessions, the line, their least-squares approximation above 30 dB (*slope*, $\alpha = 0.33$). Below 30 dB, the line has been fitted by eye. The crosses and asterisks indicate group results obtained by AME and AMP on 12 observers for the same 1,000-Hz tone bursts in a preceding experiment. Reproduced from Zwislocki (1983) with permission from the Psychonomic Society

computations were based on the two-stage model of Atteneave (1962) in which a nonlinear relationship between numbers and their subjective magnitudes is allowed.

Let us assume on the basis of prevalent experience that an observer assigns numbers to subjective magnitudes according to the power function

$$n = k\phi^{\alpha} \tag{1.19}$$

where n indicates the number, ϕ, the magnitude of the physical stimulus, α, the power exponent, and k, a dimensional constant. Let us assume further, in agreement with the experiment on line-length summation, that the subjective magnitude follows a power function of the form

$$x = b\phi^{\theta} \tag{1.20}$$

where x stands for the subjective magnitude, θ, for the power exponent, and b is a dimensional constant. If, finally, the relationship between subjective magnitudes and the numbers assigned to them is also governed by a power function, as suggested by

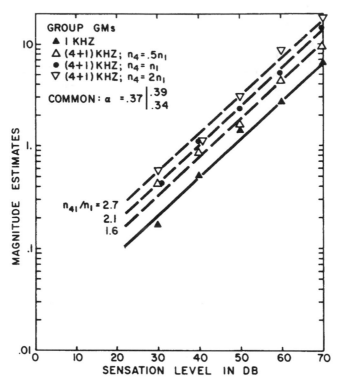

Fig. 1.35 AME of monaural pairs of tone bursts having the tone frequencies of 4,000 and 1,000 Hz and three different loudness ratios – 0.5, 1, and 2 according to Fig. 1.34 and 0.6, 1.1, 1.7 empirical. The intermittent lines are least-squares regression lines with a common slope for all the loudness ratios. The triangles and solid regression line show AME results for single 1,000-Hz bursts obtained in the same session with the burst pairs. Reproduced from Zwislocki (1983) with permission from the Psychonomic Society

variations in the slope of loudness functions (e.g. Fig. 1.19), we can write

$$n = ax^\beta \tag{1.21}$$

By combining Eqs. (1.20) and (1.21), we obtain

$$n = a(b\phi^\theta)^\beta \tag{1.22}$$

the two stage equation of Attneave in a power-function form. In this equation, $\theta\beta = \alpha$ and $ab^\beta = k$ of Eq. (1.19).

In application to the tone-burst experiment described above, we can write

$$x_{41} = x_4 + x_1 \tag{1.23}$$

with x_1 standing for the loudness of the 1-kHz tone burst, x_4, for the loudness of the 4-kHz tone burst, and x_{41}, for their sum. Replacement of the loudness terms by their numerical estimates according to Eq. (1.21), leads to

$$n_{41}^{1/\beta} = n_4^{1/\beta} + n_1^{1/\beta} \tag{1.24}$$

or

$$\frac{n_{41}}{n_1} = \left[1 + \left(\frac{n_4}{n_1} \right)^{1/\beta} \right]^{\beta} \tag{1.25}$$

For some computational purposes the latter equation can be normalized with respect to $\beta = 1$. Then,

$$\frac{n_{41}}{n_1^*} = \frac{[1 + (n_4/n_1)^{1/\beta}]^{\beta}}{[1 + (n_4/n_1)]} \tag{1.26}$$

where n_1^* corresponds to $\beta = 1$. When n_1, n_4, and n_{41} are determined experimentally, Eqs. (1.25) or (1.26) make the determination of β possible. In particular,

$$\beta = \frac{\ln(n_{41}/n_1)}{\ln 2} \tag{1.27}$$

for $n_4 = n_1$.

Once the exponent, β, is computed based on the empirical values of n_1, n_4, and n_{41} introduced into Eq. (1.25) and, in particular, Eq. (1.27), the exponent, θ, relating the subjective magnitude to the stimulus magnitude, can be determined. It is given by the relationship $\theta\beta = \alpha$ that can also be written in the form $\theta = \alpha/\beta$. The obtained numerical value can be verified independent of α and β by increasing the magnitude of one of the component stimuli, e.g. $\phi_1 = \phi_{1s}$, presented singly until its loudness becomes equal to the loudness of the component stimuli presented together. In equation form,

$$x_{1s} = x_4 + x_1 \tag{1.28}$$

Application of Eq. (1.20) leads to

$$\phi_{1s}^{\theta I} = \phi_4^{\theta I} + \phi_1^{\theta I} \tag{1.29}$$

By selecting the stimuli so that their equal magnitudes produce equal loudness magnitudes, as this is exemplified in Fig. 1.34, and by making $\phi_4 = \phi_1$, Eq. (1.29) can be simplified so that it becomes

$$\phi_{1s}^{\theta I} = 2\phi_1^{\theta I} \tag{1.30}$$

Accordingly:

$$\theta_1 = \frac{\log 2}{\log(\phi_{1s}/\phi_1)} \tag{1.31}$$

and in dB notation,

$$\theta_1 = \frac{10 \log 2}{(10 \log \phi_{1s} - 10 \log \phi_1)} \tag{1.32}$$

where $10 \log \phi_{1s} - 10 \log \phi_1$ is the measured SL-difference. If loudness additivity holds, then, according to the two-stage model $\theta_1 = \theta$.

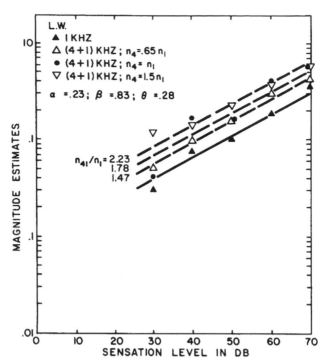

Fig. 1.36 The same as Fig. 1.35 but for an individual observer having the lowest slope of loudness growth. Reproduced from Zwislocki (1983) with permission from the Psychonomic Society

An example of individual results that depart considerably from typical data is shown in Fig. 1.36. By applying the least-squares method, the results were found to follow straight lines with an average slope of $\alpha = 0.23$ on log coordinates. The post-facto nominal ratios of numerical loudness estimates were: $n_4/n_1 = 0.65$, $n_4/n_1 = 1$ and $n_4/n_1 = 1.5$, as indicated in the figure. They are equivalent to nominal n_{41}/n_1^* ratios of 1.65, 2 and 2.5, which produced directly measured n_{41}/n_1 ratios of 1.47, 1.78 and 2.23. The quotient of the n_{41}/n_1 and n_{41}/n_1^* ratios came out to be 0.89, independent of the numerical value of n_{41}/n_1^*. Introduced into Eq. (1.26), it yielded an exponent $\beta = 0.83$, substantially different from unity. Within the group, the quotient varied between 0.86 and 1.45. The individual ratios are displayed in Table 1.2a. In general, they depended on the nominal n_4/n_1 values. For the calculation of the individual exponents, β_M, their individual means, displayed in Table 1.2b, were used. These exponents are shown in the bottom part of Table 1.2b in the β_M column together with the mean exponents, α_M, resulting from least-squares fits to the numerical loudness estimates. The table also contains the individual mean exponents θ_M obtained through division of the exponents α_M by the exponents β_M as well as the exponents α_L associated with individual line-length estimates. In the last column are shown the exponents, θ_1 obtained by applying Eq. (1.32) to individual data sets, such as exemplified graphically in Fig. 1.36. As already mentioned, calculation of θ_1 did not depend on the exponents α_M or β_M.

Table 1.2 Upper Panel: Ratios between experimental and theoretical AMEs of loudness for individual observers and their means. Lower Panel: Slopes of regression lines on double-log coordinates. Reproduced from Zwislocki (1983) with permission from the Psychonomic Society

Subject	Ratios Between Experimental and Theoretical Loudness Estimates of Individual Subjects, and Their Means [Experimental (n_{41}/n_1)/Theoretical (n_{41}/n_1)]			
	Nominal $(n_4)/n_1$			
	0.5	1.0	2.0	Mean
R.D	1.44	1.45	.89	1.26
M.T.	1.19	1.24	.93	1.12
J.F.	1.03	1.21	1.03	1.10
J.E.	1.23	.88	.86	.99
L.W.	.89	.89	.89	.89
Mean	1.16	1.13	.91	1.07

Subject	Table 2 Slopes of Log-Log Regression Lines				
	α_M	β_M	α_L	θ_M	θ_I
R.D	.56	1.33	1.33	.42	.38
M.T.	.50	1.16	1.2	.43	.43
J.F.	.25	1.10	1.0	.23	.24
L.W.	.23	.83	.88	.28	.28
Mean	.38	1.08	1.07	.34	.34

Note—α_M = *loudness estimates vs. SL;* β_M = *loudness estimates vs. loudness;* α_L = *length estimates vs. line length;* θ_M θ_I = *loudness vs. SL.*

Table 1.2b is pregnant with significant numerical constants and their relationships. The bottom row containing the group means shows from left to right that, in the presence of burst pairs, the observers tended to assign numbers to loudness magnitudes according to a somewhat steeper power function, $\alpha_M = 0.38$ on the average, than is usual for AME, as evident in Fig. 1.34; that the numbers were nearly directly proportional to loudness magnitudes, following a power function with a mean exponent, $\beta_M = 1.08$; that the numbers assigned to line length followed a power function with practically the same exponent, $\alpha_L = 1.07$; that the exponent, θ_M, derived for the relationship between the inferred loudness magnitudes and sound intensities through division of the individual α_M exponents by the β_M exponents was equal to the exponent, θ_I, derived directly on the basis of the assumed loudness summation. The common numerical value of the exponents θ_M and θ_I agrees with the slope of the apparently unbiased loudness functions of Fig. 1.34 and, more generally, with the "best" estimate of Stevens (1975), made on the basis of many direct and indirect measurements. Because Stevens based his estimate on sound pressure rather than intensity used in Table 1.2, its numerical value was doubled and amounted to 0.67.

The individual data of the table show almost perfect correlations between the numerical values of the exponents β_M and β_L as well as between the numerical values of the exponents θ_M and θ_I. The mutual consistency of all these values validates the assumption of loudness additivity on which, with the exception of α_L, they depend. In addition, the agreement between the individual numerical values of β_M and β_L confirms the direct proportionality between the subjective and the underlying physical line lengths and indicates that deviations from such proportionality in AME are due practically entirely to the deviations of the subjective magnitudes of numbers from their nominal values. Because the β_M values were derived from loudness measurements and those of α_L from subjective line length estimates, the deviations do not seem to depend on the sensory modality.

The agreement between the individual exponents β_M and β_L has a practical application. The empirically determined exponents α_L can serve as a measure of the inferred exponents β_M and be used as correction factors for the empirically determined subjective-magnitude functions. In Table 1.2a,b, the individual values β_M have been applied directly to the individual α_M values to obtain the individual θ_M values. However, the result would have been nearly the same if the α_L values were applied instead. Note that the operation was associated with a decreased interindividual variability. A similar result was obtained by Collins and Gescheider (1989) on larger populations of adults and children. Because of its theoretical and practical importance the correlation between β_M and α_L is shown graphically in Fig. 1.37. The filled circles indicate the individual data; the straight line is fitted to them by the least-squares method with a correlation coefficient of $r = 0.95$. The line approximates fairly well direct proportionality, having an angle of almost $45°$ and crossing the ordinate axis near the coordinate origin.

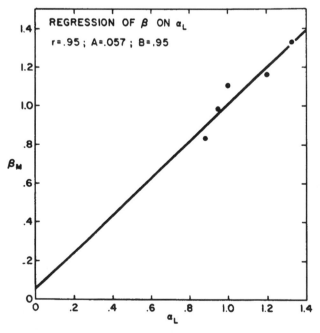

Fig. 1.37 Correlation between derived exponents β relating AME of loudness to putative true loudness magnitudes and empirical exponents α_L associated with AME of line length. The circles indicate individual data; the line is their least-squares regression line. Reproduced from Zwislocki (1983) with permission from the Psychonomic Society

Because the mean value of exponent β_M in Table 1.2b approximates unity and varies only moderately around unity in individual observers, the finding of direct proportionality between numbers and their subjective values is extended to loudness, beyond AME of subjective line length. Due to the transitivity of AME results, it must hold for other subjective magnitudes as well. The relationship between numbers and their subjective magnitudes definitely does not have a logarithmic form, as has been suggested by some theoreticians.

The demonstration of direct proportionality between numbers and their subjective magnitudes is sufficient for the validation of Stevens' power law that is generally expressed as a functional relationship between numbers assigned to subjective magnitudes and the stimulus magnitudes that evoke them. In mathematical terms Attneave's two-stage relationship in the form of Eq. (1.22), is simplified to become

$$n = ab\phi^{\theta} \tag{1.33}$$

A caveat must be inserted here, however. The simplified equation applies only to unbiased results that conserve the direct proportionality between the numbers and their subjective magnitudes $(\beta = 1)$.

1.6.1 Generality

The wide generality of the power law is evident in the results of cross-modal experiments shown graphically in Fig. 1.22. Additional results for loudness, line length, brightness, and subjective vibration intensity obtained by means of absolute scaling are displayed in Figs. 1.12–1.14, 1.20, 1.24, 1.25, 1.27, and 1.32–1.36 for groups of observers as well as for individual observers. Stevens (1975) lists power-law exponents for a large number of continua. The list is reproduced in Table 1.3. It implies that a large variety of subjective continua obey the power law. However, the exponents, although typical, cannot be considered to be completely invariant. They depend to some extent on the experimental conditions.

For completeness, examples of characteristic intensity functions for smell, taste, and warmth are added in Figs. 1.38–1.40 to those already shown in the preceding sections. The functions have been obtained by means of magnitude estimation referred to the magnitude estimate of the first stimulus presented rather than by means of AME. However, both methods give comparable results with respect to the curve slopes on log-log coordinates.

The olfactory functions (Cain, 1969), shown in Fig. 1.38, have been obtained for 5 odorants – acetone, n-propanol (C_3), n-butanol (C_4), n-pentanol (C_5) and geraniol, presented in separate sessions to 15 observers through an olfactometer that diluted the odorants in air in varying proportions. The odorants were presented for periods of 3.5 s. Under different conditions of stimulus presentation, the slopes of the functions would have been altered somewhat without changing the rank-order of their slopes. Significantly, all the data follow power functions.

Table 1.3 Exponents of power functions relating numerical estimates of subjective magnitudes to associated stimulus magnitudes, as given by Stevens. From Stevens (1975) no copyright permission granted

Continuum	Measured exponent	Stimulus condition
Loudness	0.67	Sound pressure of 3000-hertz tone
Vibration	0.95	Amplitude of 60 hertz on finger
Vibration	0.6	Amplitude of 250 hertz on finger
Brightness	0.33	5° Target in dark
Brightness	0.5	Point source
Brightness	0.5	Brief flash
Brightness	1.0	Point source briefly flashed
Lightness	1.2	Reflectance of gray papers
Visual length	1.0	Projected line
Visual area	0.7	Projected square
Redness (saturation)	1.7	Red-gray mixture
Taste	1.3	Sucrose
Taste	1.4	Salt
Taste	0.8	Saccharine
Smell	0.6	Heptane
Cold	1.0	Metal contact on arm
Warmth	1.6	Metal contact on arm
Warmth	1.3	Irradiation of skin, small area
Warmth	0.7	Irradiation of skin, large area
Discomfort, cold	1.7	Whole body irradiation
Discomfort, warm	0.7	Whole body irradiation
Thermal pain	1.0	Radiant heat on skin
Tactual roughness	1.5	Rubbing emery cloths
Tactual hardness	0.8	Squeezing rubber
Finger span	1.3	Thickness of blocks
Pressure on palm	1.1	Static force on skin
Muscle force	1.7	Static contractions
Heaviness	1.45	Lifted weights
Viscosity	0.42	Stirring silicone fluids
Electric shock	3.5	Current through fingers
Vocal effort	1.1	Vocal sound pressure
Angular acceleration	1.4	5-Second rotation
Duration	1.1	White noise stimuli

The taste functions (Moskowitz, 1970) of Fig. 1.39 have been obtained for the sweetness of sucrose in 10 experiments performed on different groups of 10 to 40 observers. In 7 experiments, the observers made 2 judgments per sucrose concentration, in 3 experiments, only one. The sucrose solutions in tap water at a temperature of 19 °C were presented to the observers in paper cups in a random order of concentrations, and the observers rinsed their mouths with tap water after every presentation. They could take as much time as they wanted before giving a response. The cluster of the data points on the left of the graph contains the unnormalized response medians of the 10 experiments. In the cluster to the right, the medians were normalized to eliminate the global inter-experiment differences. Up

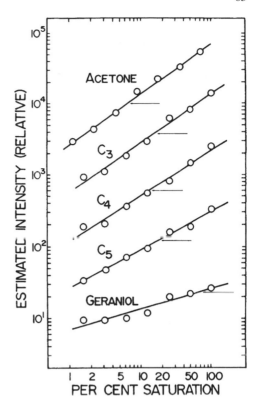

Fig. 1.38 Odor intensity measured on 15 observers by magnitude estimation as a function of odorant concentration for 5 odorants – acetone, *n*-propanol, *n*-butanol, *n*-pentanol and geraniol. For clarity, the functions were staggered along the ordinate axis. Reproduced from Cain (1969) with permission from the Psychonomic Society

to a molar concentration of about 0.5, the data points of both clusters conform to a straight-line trend on log-log coordinates and have been approximated with the help of the least-squares method by power functions having an exponent of 1.3. At higher concentrations, a saturation effect is evident, however, implying a limit on the power law. Interestingly, the exponent of 1.3 is not specific to sucrose but appears to hold for almost all sugars. Sour-tasting and bitter-tasting substances produced power functions with exponents depending somewhat on the substance but hovering around unity (Moskowitz, 1971a). The saltiness taste of sodium chloride followed a power function with an exponent similar to that for sucrose.

The sensation of warmth is of particular interest because its threshold lies well above the null of physical heat energy. The situation is somewhat similar to that of partial auditory masking by random noise that can shift the threshold of audibility of a pure tone by comparable amounts. As shown in Fig. 1.16, the slope of the loudness function becomes steeper than in the absence of the noise and the steepness increases with the threshold shift. One example of the growth of subjective warmth magnitudes with heat intensity is shown in Fig. 1.40 (J.C. Stevens and Marks, 1971). The stimulus consisted of a heat flux generated by a 1kW projection lamp aimed at the forehead. The forehead was painted with India ink to facilitate heat absorption. The variable area of exposure was controlled by aluminum masks, and the exposure

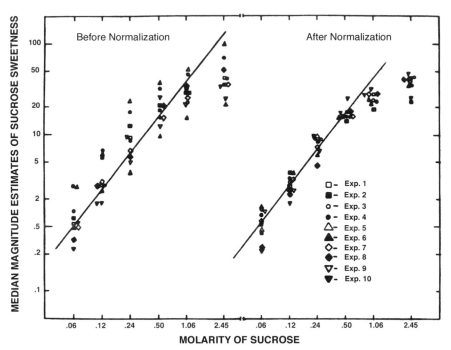

Fig. 1.39 Sweetness of sucrose diluted in various concentrations in tap water, as measured by magnitude estimation in 10 experiments on groups of 10–40 observers. Unnormalized data are shown on the left, normalized data, on the right. The straight lines were fitted to the data at the 4 lowest concentrations. Note the saturation effect at the higher concentrations. Reproduced from Moskowitz (1970) with permission from the Psychonomic Society

time, by a shutter that was opened every 30 s for 3 s. The radiant intensity was measured with a Hardy radiometer. The resulting subjective warmth data of Fig. 1.40 are based on magnitude estimates produced by 15 observers. They were fitted by a family of smooth curves obeying Eq. (1.34)

$$\psi = k(\phi - \phi_0)^\theta \tag{1.34}$$

with two assumptions – one, that the threshold of detectability, ϕ_0, decreases in direct proportion to the illuminated area and, two, that the curves must converge near the threshold of pain. Whereas, there is good evidence for the former assumption, the latter is less well justified because of the paucity of the data in the vicinity of the point of convergence. In the next chapter, a different equation and a different explanation for the systematic slope variation is suggested. Either approach produces a departure from the power law in the vicinity of the threshold, where the steepness of the curves increases on double-log coordinates. At higher intensities the law is preserved.

The examples included in this and preceding sections indicate that Stevens' Power Law holds for the relationships between psychological and physical or chemical magnitudes of many adequate stimuli, in particular when a correction is

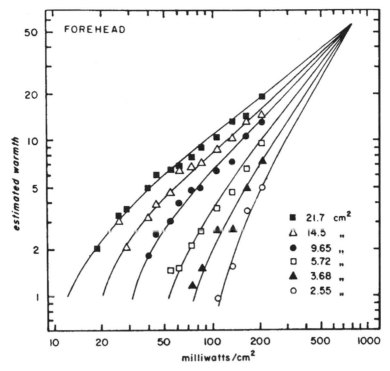

Fig. 1.40 Magnitude of warmth produced by radiant heat on the forehead and measured by magnitude estimation on 15 observers. The area of exposure served as a parameter. The data were fitted by curves obeying a power-function equation with a correction for the threshold effect (Eq. 1.34). Reproduced from Stevens and Marks (1971) with permission from the Psychonomic Society

introduced for the threshold of detectability (e.g. Marks, 1974). But the law is not universal, and whole classes of psychophysical relationships do not conform with it. The deviations are particularly evident in the sense of taste, as exemplified by sucrose in Fig. 1.39 and in hedonic scales. The latter refer to pleasantness rather than to the subjective intensity. A good example can be found in the pleasantness of the sweet taste of simple sugars, as shown in Fig. 1.41 (Moskowitz, 1971a). Whereas the sweetness magnitudes of most of these sugars tend to increase with their concentration according to power functions, the curves of pleasantness show saturation effects and even maxima beyond which they have negative slopes. Other sugars exhibit similar effects (e.g. Moskowitz, 1971b). A somewhat more extensive review of hedonic scales is provided by Marks (1974) but even every-day cooking and eating furnish ample examples of nonmonotonic pleasantness functions.

The encountered deviations from its predictions make me reluctant to accept Stevens' Power Law as *the* psychophysical law (Stevens, 1975). Nevertheless, because it applies to most intensive continua, it should be accepted as *a* psychophysical law.

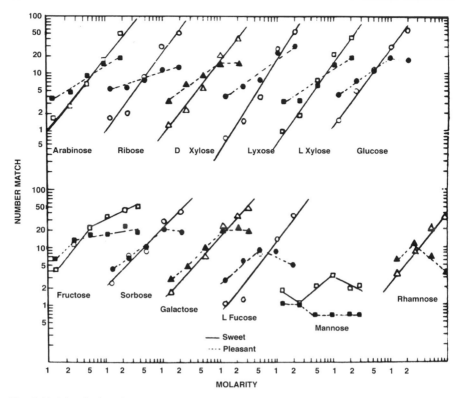

Fig. 1.41 Magnitudes of sweetness (*unfilled symbols and solid lines*) and pleasantness (*filled symbols and intermittent lines*) of 12 simple sugar solutions as functions of concentration. The data points show normalized medians obtained on 12 experienced observers by magnitude estimation without a designated reference standard. All sweetness functions, but one, conform to the power law. This is not true for the pleasantness functions. Reproduced from Moskowitz, 1971 from the American Journal do Psychology. Copyright 1971 by the Board of Trustees of the University of Illinois. Used with permission of the author and the University of Illinois Press

1.6.2 The Context Problem

Perhaps the most basic of all sciences – physics, may serve as a useful model for generation of scientific laws. Physical laws are strictly valid only under idealized conditions. Take for example the most popularly known law of Newton's: $F = a \times m$, where "F" means a force pulling an object of mass "m" that moves with an acceleration "a" as a result, and apply it to the gravitational pull of the earth.

The force of the pull, F, is directly proportional to the mass of the falling object: $F = g \times m$, where "g" is the gravitational constant. By replacing F with $g \times m$ in Newton's formula, we obtain: $g \times m = a \times m$, or $g = a$. Accordingly, all objects should fall to the earth with equal acceleration, independent of their mass. Everybody knows that this is not true under ordinary conditions – a feather falls much more slowly than a stone! Why? Because Newton's formula holds only for objects falling in vacuum. When they fall in the air under atmospheric pressure, the air resistance makes light objects fall more slowly than heavy objects.

The effect of air resistance may be used as a conceptual model for psychological context effects (Zwislocki, 1983). Loudness of a 1-kHz tone presented in quiet grows according to a power function at SPLs that exceed the threshold of audibility by a sufficient amount (e.g. Figs. 1.9, 1.10, and 1.20). When masking noise is introduced, the loudness function becomes steeper and departs from the power function, becoming concave downwards on double-logarithmic coordinates (Fig. 1.16). A similar effect is observed in vision when a luminous disc is displayed on a black surround, then the luminance of the surround is increased. An example is shown in Fig. 1.42 (J.C. Stevens, 1966). The data have been reproduced from J.C. Stevens' Ph. D. dissertation (1957) and were obtained in three experiments with 5 observers participating in each. The stimulus was produced by a Macbeth illuminometer so that the observer saw a disc subtending a visual angle of 2.4° surrounded

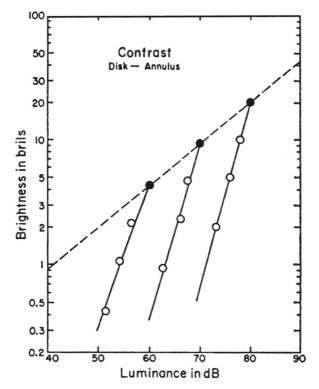

Fig. 1.42 Brightness of a disc subtending a visual angle of 2.4° surrounded by an annulus subtending a visual angle of 5.7° and having a parametrically varied luminance. The task of the 5 participating observers was to set the brightness of the disc to a given fractional value of the annulus brightness. The data points were fitted by straight lines. The intermittent line shows the brightness function obtained in the absence of surround illumination. Reproduced from Stevens (1966) with permission from the American Institute of Physics

by an annulus subtending a visual angle of 5.7°. With the annulus light turned off, the brightness of the disk grew according to a power function with an exponent of 0.33, as shown by the dashed line in the figure. When the surround was illuminated at three different luminance levels, a different one in every experiment, the brightness of the disk grew with its luminance according to a steep function until the luminance became equal to that of the surround. The brightness data were not obtained by magnitude estimation. Instead, the observers had to set the disc luminance so as to make its brightness appear 1/2, 1/4 and 1/10 as great as the surround brightness. The few data points obtained in this way were averaged by segments of straight lines on the log-log coordinates. Clearly, the resulting brightness curves, when taken over the entire luminance range, depart severely from power functions, being steep below the surround luminance and relatively flat above it.

Masking noise and surround light are stimulus contexts, although they are not usually included in discussions of context effects. They clearly point to the fact that power functions can only be approximated when context effects are minimized. They cannot be eliminated entirely in psychophysical experiments. Even in a set of test stimuli presented in the absence of other stimuli, the test stimuli provide contexts for each other. In magnitude estimation, smaller numbers tend to be assigned when the stimulus set is selected from the upper part of a possible stimulus range than when it is selected from the lower part. A prominent effect of this sort, called "contrast effect," is shown in Fig. 1.43 (Marks, 1988). The effect was obtained by

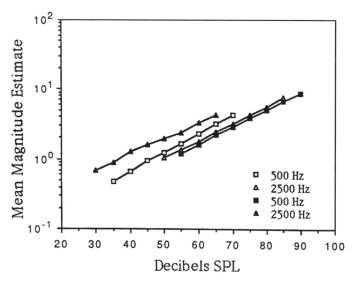

Fig. 1.43 Loudness of 500 and 2,500-Hz tone bursts having 1-s durations and presented alternately in different SPL ranges. In the first session, the 500-Hz bursts (*unfilled squares*) occupied the lower range, the 2,500-Hz burst (*unfilled triangles*), the higher range. In the second session, the ranges were reversed. The data were obtained on 16 observers by magnitude estimation without designated reference standards and geometrically averaged. Reproduced from Marks (1988) with permission from the Psychonomic Society

presenting alternately tone bursts of two different sound frequencies – 500 and 2,500 Hz, in the same sessions. In the first session, the 500-Hz tone was in a relatively low range of SPLs, and the 2,500-Hz tone in a relatively high range (*unfilled symbols*). In the next session, the levels were reversed (*filled symbols*). Clearly, the relatively higher SPL ranges produced relatively lower numerical estimates, the relatively lower SPLs, higher estimates. The large size of the contrast effect was probably caused by mixing two different stimuli in the same sessions and by implied reference standards. In AME, when tones of a single sound frequency are used, the effect may be vanishingly small (e.g. Fig. 1.12).

An opposite effect occurs when, within a test set, a weaker stimulus follows a stronger one – its subjective magnitude tends to be overestimated. When a stronger stimulus follows a weaker one, the reverse tends to be true. This is called an "assimilation effect." Because of this and related effects, randomized presentation of stimulus magnitudes is essential when minimally biased empirical functions are to be obtained.

A still different context effect results from the extent of a stimulus set. A set with a narrow magnitude range tends to produce a steeper function than a stimulus set with a broad range. The effect is particularly pronounced for very narrow ranges that, as a consequence, should be avoided. A good short overview of these effects is given by Gescheider (1997).

There is no doubt that context effects exist. The essential question is this, however: should a law, whether physical or psychological, be abandoned because it is perturbed by the presence of contexts? On the basis of centuries of experience, physicists say: no. Should psychologists not follow their example? A law constitutes a backbone on which studies of many effects can be anchored. Only because we know what happens to an object falling in vacuum, can we study the effect of air friction. Only when a psychophysical law is established in the presence of minimum context effects, can these effects be studied when more appreciable.

1.6.3 Physiological Correlates

A psychophysical law must be rooted in physiological processes, and we should be able to find its physiological correlates. Stevens (1975) seemed to think that the nonlinear transformations underlying the Power Law took place in sensory transducers – the sensory receptor cells. Because such transformations are usually compressive, they would decrease the magnitude range the rest of the system has to process. Increasing empirical evidence indicates that Stevens was right in principle but not in specifics. The transformation is not limited to the transducers and involves such peripheral organs as the mechanical parts of the cochlea in the ear and the pupil in the eye as two prominent examples.

In the ear, even the vibration amplitude of the basilar membrane is compressed, and the compression is strong enough to account for the exponent of $\sim 1/3$ of the loudness power function (e.g. Ruggero et al., 1997). The compression is preserved in

Fig. 1.44 Alternating receptor potentials of 6 inner hair cells found in 6 Mongolian gerbils and recorded intracellularly as functions of SPL. The upper panel shows peak responses, the lower panel, the response areas (integrated response as a function of tone frequency). The data of individual gerbils were normalized relative to the value at 30 dB. The curve in each panel indicates a typical human loudness function (Hellman and Zwislocki, 1961). Reprinted from Chatterjee and Zwislocki (1998) with permission from Elsevier

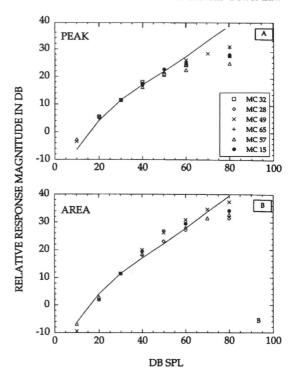

the inner hair cells, as illustrated in Fig. 1.44 by the magnitudes of their alternating receptor potentials (Chatterjee and Zwislocki, 1998). The magnitudes recorded in single cells of 6 Mongolian gerbils and marked by the various symbols have been normalized and are plotted in dB as functions of SPL. The upper panel refers to peak responses in the sound-frequency range of about 1–3 kHz, the lower panel, to response magnitudes integrated over the response area of every cell, the response area being defined by the response magnitude as a function of sound frequency. The solid lines indicate a typical loudness function (Hellman and Zwislocki, 1961). The peak response follows the loudness function up to an SPL of about 50 dB, the integrated response, up to that of about 70 dB. Clearly, loudness must be coded in terms of an aggregate of the responding cells with staggered sensitivities.

The conclusion is confirmed by the measured response of the whole auditory nerve innervating the inner hair cells. Unfortunately, meaningful measurement of such a response is difficult because of the biphasic nature of the recorded neuronal action potentials and the dispersion of response latencies of the neurons involved. Perhaps the closest approximation to the whole-nerve response was obtained by Teas et al. (1962) who used 1.05-msec acoustic impulses as stimuli and suppressed the effects of frequency sidebands by noise masking. Their normalized results, marked by filled circles, are compared in Fig. 1.45 to a typical loudness function obtained at 1 kHz and indicated by the solid line (Zwislocki, 1974). Within the SPL-range of their measurements extending over approximately 60-dB, the

Fig. 1.45 Typical loudness function at 1,000 Hz (solid curve; Hellman and Zwislocki, 1961) and related neurophysiological responses: crosses – firing rate of a typical neuron of the auditory nerve in response to a pure tone (Zwislocki, 1974); filled circles – whole-nerve response to 1-msec clicks (Teas et al. 1962); unfilled circles – stapedius-muscle response to 2000-Hz 20-msec tone bursts (Zwislocki and Shepherd, 1972; cit. Zwislocki, 2002). From Zwislocki (1974) reprinted with permission from copyright holder

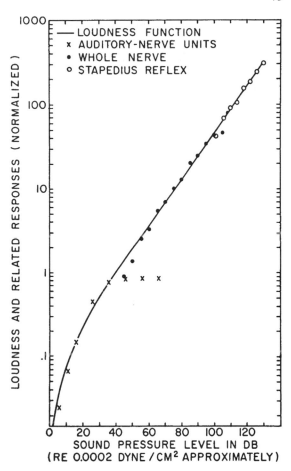

nerve response coincides with the loudness function, except at the lowest and highest SPLs. This agreement is remarkable in view of the experimental difficulties involved.

Near the threshold of detectability, only the most sensitive neurons are likely to be excited, and the auditory information is likely to be transmitted by only a few units with similar response characteristics. These characteristics are approximated by that of a typical neuron reproduced by means of the crosses in Fig. 1.45. It follows the loudness function up to about 40 dB and flattens out at higher SPLs. The saturation further confirms the conclusion that loudness is coded by an aggregate of neurons.

Integration of the responses of auditory-nerve neurons by a recording electrode is not natural and may poorly reflect the natural integration. Such an integration takes place in the stapedius muscle, however, and can be measured indirectly with the help of the acoustic-impedance change produced at the tympanic membrane by the stapedius-muscle contraction (e.g. Zwislocki, 2002, 2003). The change has been demonstrated to be directly proportional to the electro-myographic response of the muscle. The contraction of the muscle is bilateral so that it can be measured in one ear while stimulating the contralateral ear. Because it does not affect sound transmission at frequencies exceeding about 2 kHz, it can be studied in an "open-loop" condition. The impedance change itself is measured at a low sound frequency and a SPL low enough not to elicit a muscle contraction. Interestingly, only the intensity characteristic of the contralateral muscle reflex parallels the loudness function, as indicated by the unfilled circles of Fig. 1.45. The ipsilateral response grows much faster with SPL. The somewhat enigmatic phenomenon is illustrated in Fig. 1.46 by the filled symbols belonging to three subjects and two solid straight lines representing their least-squares approximations. The lines show that the rate of growth increases with stimulus duration but the power-function characteristic is maintained throughout. For comparison, the open symbols belonging to 5 subjects and their straight-line approximation show the intensity characteristic of the contralateral reflex.

The pupillary reflex of the eye may be regarded as a counterpart of the stapedius-muscle reflex in the ear. It too can be used as a natural integrator of the neural flux in the optic nerve. Complicating contrast phenomena can be avoided when ganzfeld illumination is used. Such an illumination can be achieved by covering the eye with a light-diffusing half ping-pong ball. The pupil reflex has the advantage

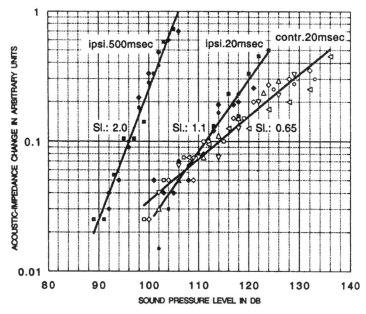

Fig. 1.46 Ipsilateral (*filled symbols*, 3 subjects) and contralateral (*unfilled symbols*, 5 subjects) responses of the human stapedius muscle to 20-msec and 500-msec tone bursts. The individual data were normalized to grand-mean values and approximated by straight-line functions on the log-log coordinates. "Sl" is an abbreviation for "slope." Reproduced from Zwislocki (2003) with permission from the National Academy of Sciences, USA

over the stapedius-muscle reflex in that it responds to faint light near the threshold of detectability and increases monotonically over an intensity range almost as great as that of brightness. It also reflects retinal abnormalities in a way resembling their effects on brightness. The similarity between the magnitude characteristics of brightness and of the pupil diameter were studied in two experimental series on the same three groups of observers. One group of 10 observers had normal vision, one group of 5 observers had retinas with missing rods (rod monochromats), the third of 11 observers, retinas with missing cones (retinitis pigmentosa). The stimuli consisted of 1-s light flashes. In the first experimental series, the observers estimated the brightness by AME, in the second, their pupil diameters were measured. Because the pupil reflex is consensual, this could be accomplished by exposing only one eye to visible light and measuring the pupil diameter in the other eye by infrared photography (Barlow, 1980). The results of brightness measurements are summarized in Fig. 1.47, those of the change in pupil diameter, in Fig. 1.48. Note the similarity of functional relationships, except for the saturation of the pupillary reflex above −5 log Lamberts of light intensity in the population with normal vision. However, below −5 log Lamberts, the log slope of the change of the pupil-diameter is about 1/2 that of the brightness slope. The slopes would be equal, had Barlow plotted the pupil area instead of its diameter. For rod monochromats, both the brightness and the pupil-diameter curves are essentially flat above −5 log Lamberts. Interestingly, at low-to-moderate light intensities, they lie above the curves for observers with normal vision. For observers with retinitis pigmentosa, the threshold response

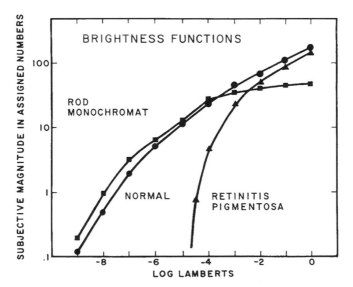

Fig. 1.47 Brightness of 1-msec light flashes in ganzfeld illumination measured as a function of luminance by AME on 10 observers with normal vision, 5 observers with missing rods (rod monochromats) and 11 observers with missing cones (retinitis pigmentosa). The curves are drawn through the group medians. Reproduced from Barlow (1980) with permission from S. Karger, AG, Basel

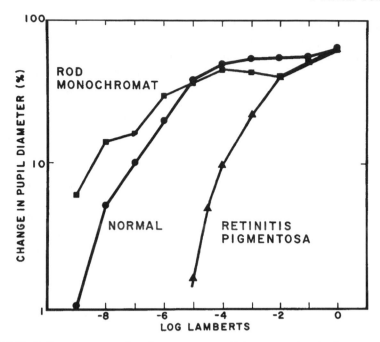

Fig. 1.48 Pupil diameter measured on the same groups of observers and under the same stimulus conditions as in Fig. 1.47. Reproduced from Bolanowski et al. (1991)

is moved from about −9 to about −5 log Lamberts, and the curves are steepened so that they converge on the normal curves at high light intensities. Note again that the pupil-diameter curves are about half as steep as the brightness curves.

Another opportunity for studying a neurophysiological correlate of the Power Law is afforded by the chorda tympani nerve innervating the anterior part of the tongue and responding to taste substances, such as salt (NaCl), sucrose and citric acid. Due to an anatomical peculiarity, the nerve curses through the middle ear cavity and can be accessed in humans during middle-ear surgery by means of an electrode. This is particularly easy in cases of otosclerosis in which the tympanic membrane is resected and pushed aside. The response of the whole nerve appears to be close to a simple sum of the component fiber responses, as can be deduced from the shape of the action potentials that are nearly monophasic, and from the similarity of its intensity characteristic to that of the fibers (for reviews, Zotterman, 1971; Sato, 1971). When, before surgery, patients estimate numerically taste magnitudes of the substances involved, a direct comparison of the psychophysical and neurophysiological responses can be made. Two examples obtained on two patients are given in Fig. 1.49, one for citric acid, the other for sucrose (Borg et al., 1967). Both substances were placed on the tongue by means of a special plastic applicator with a 15-ml reservoir. In the figure, the crosses indicate the mean numerical estimates, the circles – the peak nerve responses. Good agreement between them is evident. The numerical estimates for sucrose are also in excellent agreement with those of

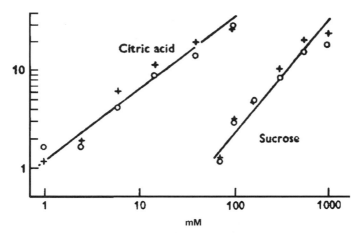

Fig. 1.49 Psychophysical (*crosses*) and whole-nerve (*circles*) responses of the corda-tympani nerve to citric acid and sucrose placed on the tongue in two patients undergoing middle-ear surgery. The data points indicate the mean responses of the two patients, the straight lines, the power-function fits to the data. Reproduced from Borg et al. (1967) with permission from Blackwell Publishers

Fig. 1.39 obtained by Moskowitz (1970). Note especially that the psychophysical as well as the neurophysiological data for citric acid follow a power function with an exponent on the order of 0.6, and those for sucrose, one with an exponent of about 1.0. Note also a saturation effect in all the sucrose data at high concentrations.

The various examples of congruence between the psychophysical and physiological data given above indicate unambiguously that Stevens' psychophysical power law applied to sensory systems is rooted in peripheral mechanisms and that the non-linear compressive transformation, if any, takes place as distally as is structurally possible. In this way, proximal parts of the systems do not have to operate over magnitude ranges equal to the dynamic ranges of stimulus intensities that are enormous in some sense modalities.

References

Atteneave, F. Perception and related areas. In S. Koch (Ed.), *Psychology: A Study of a Science 4* (pp. 619–659). New York: McGraw-Hill, 1962.

Barlow, R.B. Brightness sensation and pupil reflex in normals, rod monochromats, and patients with retinitis pigmentosa. *Adv. Ophthalmol.* 41: 149–216, 1980.

Barlow, R.B., and Verrillo, R.T. Brightness sensation in a ganzfeld. *Vision Res.* 16: 1291–1297, 1975.

Bernoulli, D. Exposition of a new theory on the measurement of risk. Originally published in Latin in 1738. Translation in *Econometrica* 22, 23–35, 1954.

Bolanowski, S.J., Jr., Zwislocki, J.J., and Gescheider, G.A. *Ratio Scaling of Psychological Magnitude: In Honor of the Memory of S.S. Stevens.* Hillsdale, NJ: Lawrence Erlbaum Associates, 1991.

Borg, G., Diamant, H., Ström, L., and Zotterman, Y. The relation between neural and perceptual intensity: A comparative study on the neural and psychophysical response to taste stimuli. *J. Physiol.* 192: 13–20, 1967.

Cain, W.S. Odor intensity: Differences in the exponent of the psychophysical function. *Percept. Psychophys.* 6(6a): 349–354, 1969.

Chatterjee, M., and Zwislocki, J.J. Cochlear mechanisms of frequency and intensity coding. II. Dynamic range and the code for loudness. *Hear. Res.* 124: 170–181, 1998.

Churcher, B.G. A loudness scale for industrial noise measurement. *J. Acoust. Soc. Am.* 6: 216–226, 1935.

Collins, A.A., and Gescheider, G.A. The measurement of loudness in children and adults by absolute magnitude estimation and cross-modality matching. *J. Acoust. Soc. Am.* 85: 2012–2021, 1989.

Delboeuf, J. *Etude psychologique. Recherches théoriques et expérimentales sur la mesure des sensations et spécialement des sensations de lumière et de fatigue.* Brussels, 1873.

Ekman, G. Is the power law a special case of Fechner's law? *Percept. Mot. Skills* 19: 730, 1964.

Fechner, G.T. *Elemente der Psychophysik*, 1860. Vol. I available in English translation as *Elements of Psychophysics*. New York: Holt, Rinehart and Winston, 1966.

Fletcher, H. Auditory patterns. *Rev. Mod. Physics* 12: 47–65, 1940.

Fletcher, H., and Munson, W.A. Loudness, its definition, measurement and calculation. *J. Acoust. Soc. Am.* 5, 82–108, 1933.

Gescheider, G.A. *Psychophysics: The Fundamentals (Third Edition).* Mahwah, NJ: Lawrence Erlbaum Associates, 1997.

Gescheider, G.A., and Hughson, B.A. Stimulus context and absolute magnitude estimation: A study of individual differences. *Percept. Psychophys.* 50: 45–57, 1991.

Hartline, H.K., and Graham, C.H. Nerve impulses from single receptors in the eye. *J. Cell. Comp. Physiol.* 1: 277–295, 1932.

Hellman, R.P. *On some factors affecting loudness as a function of intensity.* Master's thesis, Syracuse University, Syracuse, New York, 1960.

Hellman, R.P., and Meiselman, C.H. Prediction of individual loudness functions from cross-modality matching. *J. Speech Hear. Res.* 31: 605–615, 1988.

Hellman, R.P., and Meiselman, C.H. Loudness relations for individuals and groups in normal and impaired hearing. *J. Acoust. Soc. Am.* 88(6): 2596–2606, 1990.

Hellman, R.P., and Meiselman, C.H. Rate of loudness growth for pure tones in normal and impaired hearing. *J. Acoust. Soc. Am.* 93(2): 966–975, 1993.

Hellman, R.P., and Zwislocki, J.J. Some factors affecting the estimation of loudness. *J. Acoust. Soc. Am.* 33(5): 687–694, 1961.

Hellman, R.P., and Zwislocki, J.J. Monaural loudness function T 1000 cps and interaural summation. *J. Acoust. Soc. Am.* 35(6): 856–865, 1963.

Hellman, R.P., and Zwislocki, J.J. Loudness function of a 1000-cps tone in the presence of a masking noise. *J. Acoust. Soc. Am.* 36(9): 1618–1627, 1964.

Hellman, R.P., and Zwislocki, J.J. Loudness determination at low sound frequencies. *J. Acoust. Soc. Am.* 43(1): 60–64, 1968.

James, W. *Principles of Psychology.* New York: Holt, 1890.

Jastrow, J. The psycho-physic law and star magnitude. *Am. J. Psychol.* 1: 112–127, 1887.

Levelt, W.J.M., Riemersma, J.B., and Bunt, A. Binaural additivity of loudness. *Brit. J. Math. Stat. Psychol.* 25: 51–68, 1972.

Luce, R.D., and Tukey, J.W. Simultaneous conjoint measurement: A new type of fundamental measurement. *J. Math. Psychol.* 1: 1–27, 1964.

MacKay, D.M. Psychophysics of perceived intensity: A theoretical basis for Fechner's and Stevens' laws. *Science* 139: 1213–1216, 1963.

Mansfield, R.J.W. Brightness function: Effects of area and duration. *J. Opt. Soc. Am.* 63: 913–920, 1973.

Marks, L.E. *Sensory Processes: The New Psychophysics.* New York: Academic Press, 1974.

Marks, L.E. Binaural summation of the loudness of pure tones. *J. Acoust. Soc. Am.* 64: 107–113, 1978.

Marks, L.E. Magnitude estimation and sensory matching. *Percept. Psychophys.* 43(6): 511–525, 1988.

Marks, L.E. Reliability of magnitude matching. *Percept. Psychophys.* 49: 31–37, 1991.

Matthews, B.H.C. The response of a single end organ. *J. Physiol.* 71: 64–110, 1931.

McKay, D.M. Psychophysics of perceived intensity: A theoretical basis for Fechner's and Steven's laws. *Science* 139: 1213–1216, 1963.

Moskowitz, H.R. Ratio scales of sugar sweetness. *Percept. Psychophys.* 7(5): 315–320, 1970.

Moskowitz, H.R. Intensity scales for pure tastes and for taste mixtures. *Percept. Psychophys.* 9(1a): 51–56, 1971a.

Moskowitz, H.R. The sweetness and pleasantness of sugars. *Am. J. Psychol.* 84(3): 387–405, 1971b.

Newman, E.B. The validity of the just noticeable difference as a unit of psychological magnitude. *Trans. Kans. Acad. Sci.* 36: 172–175, 1933.

Richardson, L.F., and Ross, J.S. Loudness and telephone current. *J. Gen. Psychol.* 3: 288–306, 1930.

Ruggero, M.A., Rich, N.C., Recio, A., Narayan, S.S., and Robles, L. Basilar-membrane responses to tones at the base of the chinchilla cochlea. *J. Acoust. Soc. Am.* 101(4): 2151–2163, 1997.

Sato, M. Neural coding in taste as seen from recordings from peripheral receptors and nerves. In L. M. Beidler (Ed.), *Handbook of Sensory Physiology: Chemical Senses, Taste* (Vol. 2, pp. 116–147). Berlin: Springer-Verlag, 1971.

Schmidt, J.M., and Smith, J.J.B. Short interval time measurement by a parasitoid wasp. *Science*, 237: 903–905, 1987.

Shepard, R.N. On the status of "direct" psychological measurement. In C.W. Savage (Ed.), *Minnesota Studies in the Philosophy of Science* (Vol. 9, pp. 441–490). Minneapolis: University of Minnesota Press, 1978.

Stevens, J.C. A comparison of ratio scales for the loudness of white noise and the brightness of white light. Doctoral dissertation, Harvard University, 1957.

Stevens, J.C., and Mack, J.D. Scales of apparent force. *J. Exp. Psychol.* 58: 405–413, 1959.

Stevens, J.C., and Marks, L.E. Spatial summation and the dynamics of warmth sensation. *Percept. Psychophys.* 9: 291–298, 1971.

Stevens, J.C., Mack, J.D., and Stevens, S.S. Growth of sensation on seven continua as measured by force of handgrip. *J. Exp. Psychol.* 59: 60–67, 1960.

Stevens, S.S. A scale for the measurement of a psychological magnitude: Loudness. *Psychol. Rev.* 43: 405–416, 1936.

Stevens, S.S. On the theory of scales of measurement. *Science* 103: 677–680, 1946.

Stevens, S.S. *Handbook of Experimental Psychology*. New York: Wiley, 1951.

Stevens, S.S. On the brightness of lights and the loudness of sounds [Abstract]. *Science* 118: 576, 1953.

Stevens, S.S. The measurement of loudness. *J. Acoust. Soc. Am.* 27: 815–820, 1955.

Stevens, S.S. The direct estimation of sensory magnitudes-loudness. *Am. J. Psychol.* 69:1–25, 1956.

Stevens, S.S. Problems and methods of psychophysics. *Psychol. Bull.* 55: 177–196, 1958.

Stevens, S.S. Cross-modality validations of subjective scales for loudness, vibrations, and electric shock. *J. Exp. Psychol.* 57: 201–209, 1959.

Stevens, S.S. On the new psychophysics. *Scand. J. Psychol.* 1: 27–35, 1960.

Stevens, S.S. Matching functions between loudness and ten other continua. *Percept. Psychophys.* 1: 5–8, 1966.

Stevens, S.S. *Psychophysics: Introduction to its Perceptual, Neural, and Social Prospects*. New York: Wiley & Sons, 1975.

Stevens, S.S., and Greenbaum, H.B. Regression effect in psychophysical judgment. *Percept. Psychophys.* 1: 439–446, 1966.

Stevens, S.S., and Guirao, M. Loudness, reciprocity, and partition scales. *J. Acoust. Soc. Am.* 34: 1466–1471, 1962.

Teas, D.C., Eldredge, D.H., and Davis, H. Cochlear responses to acoustic transients: An interpretation of whole-nerve action potentials. *J. Acoust. Soc. Am.* 34: 1438–1459, 1962.

Teghtsoonian, M., and Beckwith, J.B. Children's size judgments when size and distance vary. Is there a developmental trend to overconsistency? *J. Exp. Child Psychol.* 22: 23–39, 1976.

Treisman, M. Sensory scaling and the psychophysical law. *Q. J. Exp. Psychol.* 16: 11–22, 1964.

Verrillo, R.T. Effect of contractor area in the vibrotactile threshold. *J. Acoust. Soc. Am.* 35: 1962–1966, 1963.

Verrillo, R.T. Comparison of vibrotactile threshold and suprathreshold responses in men and women. *Percept. Psychophys.* 26: 20–24, 1979a.

Verrillo, R.T. Change in vibrotactile thresholds as a function of age. *Sens. Processes* 3: 49–59, 1979b.

Verrillo, R.T. Absolute estimation of line length in three age groups. *J. Gerontol.* 36:625–627, 1981.

Verrillo, R.T. Stability of line-length estimates using the method of absolute magnitude estimation. *Percept. Psychophys.* 33: 261–265, 1983.

Verrillo, R.T., Fraioli, A., and Smith, R.L. Sensory magnitude of vibrotactile stimuli. *Percept. Psychophys.* 6: 366–372, 1969.

Verrillo, R.T., and Irvin, G. Absolute estimation of line length as a function of orientation and contrast polarity. *Sens. Processes* 3: 261–274, 1979.

Weber, E.H. *De pulsu, resorptione, auditu et tactu: Annotationes anatomicae et physiologicae.* Leipzig: Koehler, 1834.

Zinnes, J.L. Scaling. *Annu. Rev. Psychol.* 20: 447–478, 1969.

Zotterman, Y. The recording of the neural response from human taste nerves. In L.M. Beidler (Ed.), *Handbook of Sensory Physiology: Chemical Senses, Taste* (Vol. 2, pp. 102–115). Berlin: Springer-Verlag, 1971.

Zwislocki, J.J. A power function for sensory receptors. In H.R. Moskowitz, B. Scharf, & J.C. Stevens (Eds.), *Sensation and Measurements.* Dordrecht, Holland: Reidel, 1974.

Zwislocki, J.J. Group and individual relations between sensation magnitudes and their numerical estimates. *Percept. Psychophys.* 33: 460–468, 1983.

Zwislocki, J.J. Natural Measurement. In S.J. Bolanowski, Jr. & G.A. Gescheider (Eds.), *Ratio Scaling of Psychological Magnitude: In Honor of the Memory of S.S. Stevens* (pp. 19–26). Hillsdale, NJ: Lawrence Erlbaum Associates, 1991.

Zwislocki, J.J. Auditory system: Peripheral nonlinearity and central additivity, as revealed in the human stapedius-muscle reflex. *Proc. Natl. Acad. Sci. USA* 99(22): 14601–14606, 2002.

Zwislocki, J.J. A look at neural integration in the human auditory system through the stapedius muscle reflex. *Proc. Nat. Acad. Sci. USA* 100(5): 9073–9078, 2003.

Zwislocki, J.J., and Goodman, D.A. Absolute scaling of sensory magnitudes: A validation. *Percept. Psychophys.* 28(1): 28–38, 1980.

Zwislocki, J.J., Damianopoulos, E.N., Buining, E., and Glantz, J. Central masking: Some steady-states and transient effects. *Percept. Psychophys.* 2: 59–64, 1967.

Chapter 2
Law of Asymptotic Linearity

2.1 Definition and Genesis

In this chapter, a new law is proposed that, to the best of my knowledge, has not yet been discussed as such in the literature. According to the law, all subjective magnitudes grow linearly with the intensities of the stimuli that evoke them near their thresholds of detectability. The relationship was first discovered during auditory measurements concerning Stevens' Power Law and is described here initially for loudness, then generalized. When sufficiently small stimulus magnitudes were included, the resulting loudness curves deviated from the power law and, on double-logarithmic coordinates, bent downward, becoming gradually steeper. A typical example is shown in Fig. 2.1 where loudness magnitudes of a 1,000-Hz tone are plotted over sound-intensity abscissas expressed in dB (Hellman and Zwislocki, 1963). The solid curve has been determined by the method of magnitude estimation based on two reference standards, as described in the preceding chapter (Hellman and Zwislocki, 1961). The slanted crosses show averages of 12 studies computed by Robinson (1957), in which various methods were used. Filled circles indicate the data of Stevens (1956) obtained by magnitude estimation with the reference standards chosen by the observers themselves; open symbols and filled triangles, the data of Scharf and Stevens (1961) obtained by magnitude estimation with a designated reference standard and by halving and doubling; the vertical crosses, the data determined by Feldtkeller et al. (1959) with the help of the same method. The excellent agreement between the various sets of data and the curve suggests that the curve accurately represents the loudness of a 1,000-Hz tone as a function of its intensity. Of particular interest is the asymptotic convergence of the curve on a linear relationship between loudness and sound intensity near the threshold of audibility, as indicated by the straight line having the coordinates of 0.01 at zero SL (threshold of audibility) and 1 at 20 dB.

The linear relationship between the loudness of a 1,000-Hz tone and its sound intensity near the threshold of audibility may have been first noticed by Zwicker and Feldtkeller (1956). Their graphical representation of the relationship is shown

J.J. Zwislocki, *Sensory Neuroscience: Four Laws of Psychophysics,*
DOI: 10.1007/978-0-387-84849-5_2,
© Springer Science+Business Media LLC 2009

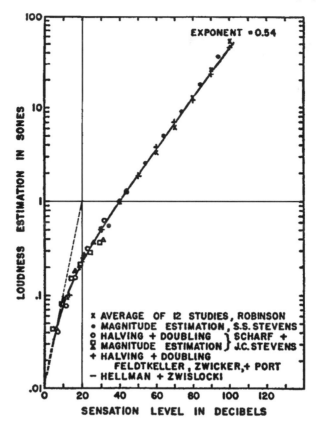

Fig. 2.1 A typical binaural loudness function (solid line) determined with two reference standards and compared to the results of five other studies performed with different methods. The intermittent straight line shows a linear relationship between loudness and sound intensity. Modified from Hellman and Zwislocki (1963), reproduced with permission from the American Institute of Physics

in Fig. 2.2, where it is denoted symbolically as $N \sim p^2$ with N standing for loudness and p for sound pressure. Of course, sound intensity is directly proportional to the square of the latter.

Is the linear relationship between loudness and sound intensity near the threshold of audibility an exclusive property of the 1,000-Hz tone, or does it extend to other sound frequencies? An experiment in which loudness was determined by the method of numerical magnitude balance at 100 Hz, 250 Hz and 1,000 Hz proved the latter to be true (Hellman and Zwislocki, 1968). As an example, geometric-mean data obtained on nine observers at 100 Hz are shown in Fig. 2.3 on double-logarithmic coordinates. The filled circles resulted from magnitude estimation, the crosses, from magnitude production. The solid line joins the geometric means of their interpolated values. The straight line with the coordinates 40 dB, 0.001 dB and 60 dB, 0.1 indicates a linear relationship between loudness and sound pressure squared. It parallels the mean loudness curve at its lowest values. Interestingly, the

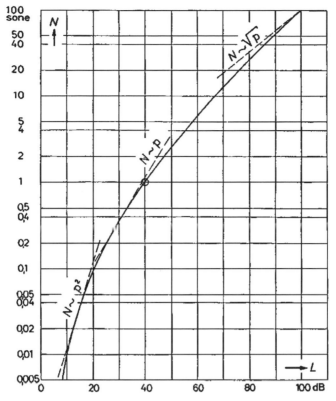

Fig. 2.2 Loudness as a function of SPL (L) determined on eight listeners by the method of halving and doubling. Loudness is indicated by N and sound pressure by p. The intermittent tangent lines indicate the slope of the function at three locations. The lowest parallels a linear relationship between loudness and sound pressure squared. Reproduced from Zwicker and Feldtkeller (1956) with permission from S. Hirzel Verlag

curve becomes somewhat steeper above these values before flattening. Geometric-mean curves obtained at all three sound frequencies are shown in Fig. 2.4. Slanted crosses indicate the respective thresholds of detectability. As a reference, straight lines with a slope of one with respect to the sound pressure squared and originating at 0 dB, 20 dB and 40 dB, respectively, are included in the graph. They are paralleled by all the loudness functions, irrespective of sound frequency. Consequently, the linearity of loudness functions near the threshold of detectability appears to hold for all sound frequencies at and below 1,000 Hz.

The slope constancy of the near-threshold loudness curves between 100 Hz and 1,000 Hz can be verified by direct inter-frequency loudness matching. Many studies were performed in which the loudness magnitudes of tones at various sound frequencies were matched to those at 1,000 Hz. The procedure consisted essentially of finding sound intensity levels referred to those at 1,000 Hz, which produced the same loudness magnitudes. Data obtained in this way at 100 Hz and 250 Hz in some

Fig. 2.3 Loudness of a 100-
Hz tone determined by magnitude balance as a function
of SPL. Filled circles indicate geometric means of data
obtained by absolute magnitude estimation, crosses, those
obtained by absolute mag-
nitude production, the solid
curve indicates their interpolated geometric means. The
solid straight line indicates a
linear relationship between
loudness and sound intensity.
Modified from Hellman and
Zwislocki (1968), reproduced
with permission from the
American Institute of Physics

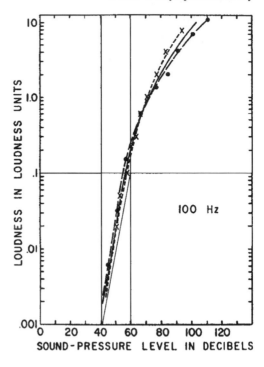

Fig. 2.4 Loudness functions determined at three
sound frequencies, 100, 250
and 1,000 Hz, by numerical
magnitude balance. Crosses
indicate the corresponding
thresholds of audibility, and
the slanted straight lines, a
linear relationship between
loudness and sound intensity.
Modified from Hellman and
Zwislocki (1968), reproduced
with permission from the
American Institute of Physics

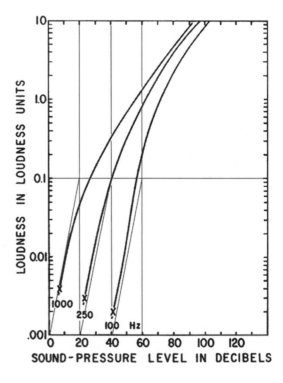

Fig. 2.5 Loudness-level curves at 100 and 250 Hz referred to the loudness at 1,000 Hz and obtained by horizontal cuts through the family of curves of Fig. 2.4. The various symbols indicate the results of corresponding direct loudness matches obtained in several studies. The slanted lines parallel the reference diagonal line and show that, at near-threshold intensities, all loudness functions are parallel to each other

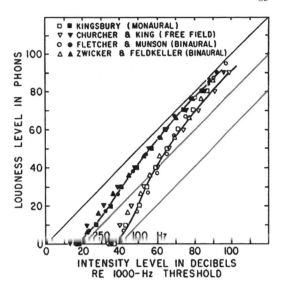

classical studies spanning a period of 40 years are shown in Fig. 2.5. They are compared to the results derived from Fig. 2.4 by sectioning horizontally the family of the curves of that figure (Hellman and Zwislocki, 1968). Note that, at near threshold levels, the curves with all the data points clustered around them parallel the diagonal line indicating the reference slope at 1,000 Hz. Similar, although less detailed, results were obtained at higher sound frequencies (e.g. Scharf, 1978). The conclusion seems warranted that, at least for pure tones, loudness is related linearly to sound intensity near the threshold of audibility.

What happens when the test tone is not presented in quiet but in the presence of masking noise that increases the slope of the loudness function, or the slope is increased by an auditory defect?

Results of a masking experiment performed on a 1,000-Hz tone are shown graphically in Fig. 2.6 (Hellman and Zwislocki, 1964). The masking stimulus consisted of an octave band of random noise centered on the test-tone frequency. It was presented at two intensity levels, so as to produce threshold shifts of 40 dB and 60 dB, respectively. The loudness levels of the partially masked tone were measured directly by comparing them to the loudness of the same tone in the absence of the masking noise. The corresponding results are shown in Fig. 2.6 by the filled circles. The abscissa axis refers to the threshold of audibility in the masked ear in the absence of the masker, the ordinate axis, to the threshold of audibility in the unmasked ear. The diagonal line corresponds to the loudness level of the unmasked tone. The curves indicate loudness levels derived from loudness measurements by the method of magnitude production. The results of these measurements agreed better with direct loudness matches than did the results of numerical loudness balance. The straight lines drawn through the threshold points parallel to the diagonal line are linearly related to sound intensity. The loudness-level curves appear to converge on them near the thresholds but the paucity of the data points leaves some uncertainty in this respect.

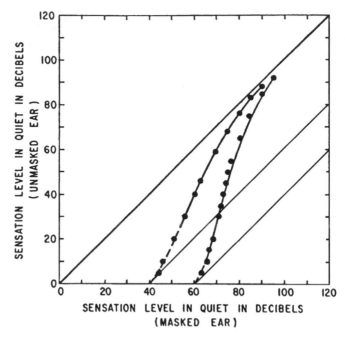

Fig. 2.6 Loudness-level curves obtained by comparing the loudness of a partially masked 1,000-Hz tone to that of an unmasked 1,000-Hz tone at two masking levels. Corresponding data derived from absolute magnitude production are indicated by filled circles. Slanted lines parallel to the diagonal indicate that masking did not alter the slope of the loudness functions near the threshold of audibility. Modified from Hellman and Zwislocki (1964), reproduced with permission from the American Institute of Physics

More convincing results were presented by B.C.J. Moore (2004) who performed experiments on four observers with sensorineural hearing loss, most likely of cochlear origin. In three observers, the hearing loss was bilaterally asymmetrical, in the fourth, it was symmetrical but affected exclusively sound frequencies above 1,000 Hz. The experiments aimed specifically at loudness growth near the threshold of audibility. They consisted of both threshold measurements and heterofrequency loudness matching. For the latter, a frequency associated with a substantial hearing loss was paired with a frequency associated with a minimal hearing loss. The experiments were performed by means of modern adaptive procedures with respect to both threshold measurements and loudness matching, assuring maximal accuracy of the results. In all four observers, when the threshold was approached, loudness grew at a rate independent of hearing loss and the slope of the loudness curve at higher sensation levels. Two examples of Moore's graphs are reproduced in Figs. 2.7 and 2.8. In both, the sensation levels corresponding to the greater hearing loss are given by the abscissas, those corresponding to the smaller hearing loss, by the ordinates. The dashed diagonal line indicates parallel loudness growth in both instances. Clearly, at near-threshold levels, all the experimental points converge on the diagonal line, indicating the same rate of loudness growth independent of hearing loss. Since the latter

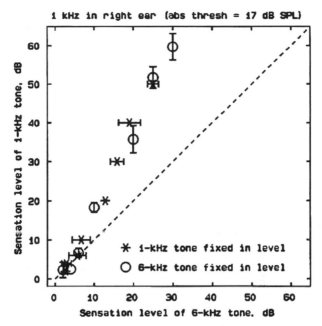

Fig. 2.7 Sensation level of a 1-kHz tone in the presence of a small hearing loss as a function of the sensation level of a 6 kHz tone in the presence of a greater hearing loss, both tones presented to the same ear at equal loudness. The slope of the resulting implied curve converges on the slope of the diagonal, indicating that the difference in hearing loss did not affect the slope of the loudness function near the threshold of audibility. Reproduced from Moore (2004) with permission from the American Institute of Physics

was negligible in one ear, the growth had to be linearly related to sound intensity, as was shown in older experiments (e.g. Hellman and Zwislocki, 1963).

The experimental evidence cited above is consistent throughout with a constant loudness growth, linearly related to sound intensity, at near threshold sensation levels, independent of sound frequency, hearing loss, or the rate of loudness growth at higher levels. The putative biophysical process responsible for the constancy is described in the following section. It is directly associated with the basic tenants of the theory of signal detectability (Green and Swets, 1966) and, in this way, provides a bridge between this theory and loudness scaling.

2.2 Underlying Biophysical Process

In all systems that are not at a temperature of absolute zero, heat is associated with molecular motion. Sensory receptors are no exception, and molecular noise was included explicitly or implicitly for many decades in the analyses of the detection of sensory signals. For example, Miller (1947) suggested that the masking effect

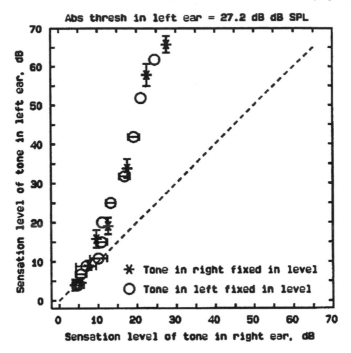

Fig. 2.8 Similar to Fig. 2.7, except that the tones were presented to different ears. Reproduced from Moore (2004) with permission from the American Institute of Physics

exerted by an extrinsic auditory noise on a pure-tone signal departed from a constant signal-to-noise ratio near the absolute threshold as a result of intrinsic noise. In vision, an analogous phenomenon is attributed to internal noise called "dark light" (e.g. Barlow, 1972). Later on, the internal noise became a fundamental postulate of the theory of signal detectability, and sensory thresholds became defined as percentages of successful discriminations between signal-plus-noise and noise-alone events (e.g. Green and Swets, 1966).

Internal noise is difficult to detect in the electrical potentials of sensory receptors because of the interfering noise of recording electrodes but its effect is clearly visible in the spontaneous activity of neurons innervating the receptors, where it appears in a digital form. The activity has been particularly thoroughly studied in afferent neurons of the auditory nerve. An example of a corresponding input/output characteristic of an auditory neuron is shown in Fig. 2.9 on double-logarithmic coordinates. The crosses indicate recorded firing rates and the star the average spontaneous activity according to Kiang (1968). The solid line through the crosses shows a theoretical approximation of the data with the inclusion of the spontaneous activity, the intermittent line, after its subtraction (Zwislocki, 1974). The lowest straight line indicates an asymptote of the intermittent line with the slope of one, in other words, a linear relationship to sound intensity.

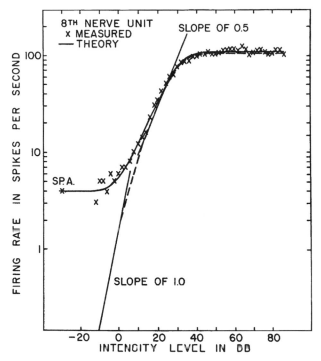

Fig. 2.9 A typical intensity characteristic of an auditory-nerve fiber. The crosses show the empirical data, the curves, their theoretical approximations. The spontaneous activity (SP. A.) is indicated by the star. The intermittent line shows the total neuronal firing rate less the spontaneous activity. It converges at low intensities on a tangent linearly related to sound intensity. Reproduced from Zwislocki (1974), with permission from copyright holder

The asymptote has been obtained analytically on the assumption that the spontaneous activity expresses the internal noise intensity that is added to the signal intensity (Zwislocki, 1973, 1974). Accordingly, the total stimulus intensity is

$$S_T = (S + N_I) \tag{2.1}$$

where S means the signal intensity and N_I, the internal noise intensity. At medium signal levels, the firing rate follows a slope of 0.5, in other words, the stimulus amplitude. This can be expressed mathematically by taking the square root of the expression in the parenthesis, so that

$$R(S_T) = A(S + N_I)^{0.5} \tag{2.2}$$

where $R(S_T)$ means the total firing rate and A is a dimensional constant. The latter equation can be written in the form

$$R(S_T) = R_0(1 + S/N_I)^{0.5} \tag{2.3}$$

with $R_o = AN_I^{0.5}$ standing for the spontaneous activity. For $S/N_I \ll 1$, Eq. 2.3 can be approximated by the first two terms of its Taylor expansion, leading to

$$R(S_T) \cong R_o(1 + 0.5 \, S/N_I) \tag{2.4}$$

By subtracting R_o, we obtain the driven neural firing rate,

$$R_D(S) \cong 0.5 \, R_o S/N_I \tag{2.5}$$

Note that it is directly proportional to the stimulus intensity, not amplitude. Equally important, the direct proportionality is independent of the exponent, 0.5 and would not be affected if the exponent were different or even the exponential function replaced by another ascending, monotonic function. This is so because of the properties of the Taylor expansion. As a consequence, all the sensory neurons exhibiting spontaneous activity should have near-threshold firing rates that are directly proportional to the stimulus intensity.

What happens when there is no spontaneous neural activity or the activity is so small that it cannot be used as a measure of the internal noise. Under such conditions, we can write for the receptor potential with the continuous use of the power-function approximation

$$E = B(E_s + E_I)^\beta. \tag{2.6}$$

where E means the total receptor potential, E_S and E_I, the receptor potentials generated by the signal and internal noise, respectively, β, a generalized power exponent, and B, a dimensional constant. For the neural firing rate, we can write on the assumption that it is linearly related to the receptor potential (e.g. Fuortes, 1971)

$$R(S_T) = C(E_S + E_I)^\beta - T \tag{2.7}$$

where $R(S_T)$ signifies the total firing rate, C, a dimensional constant and T, the firing threshold. When $E_S \ll E_I$, we can again use the Taylor approximation, so that

$$
\begin{aligned}
R(S_T &= CE_I^\beta(1 + \beta E_S/E_I) - T \\
T &\leqslant CE_I^\beta
\end{aligned}
\tag{2.8}
$$

or

$$R(S_T) = C\beta E_I^{\beta-1}E_S + (CE_I^\beta - T) \tag{2.9}$$

Accordingly, the driven firing rate remains linearly related to the signal intensity.

The above analysis shows that, for signal intensities smaller than the intensity of the internal noise, the neuronal firing rate must be approximately linearly related to the intensity of the stimulating signal. But does such linearity hold for psychological responses that are not controlled by single neurons but, rather, aggregates of neurons? One necessary condition is satisfied – the sum of linear functions is a linear function. More specifically, if the firing rates of neurons in an aggregate follow linear functions, the same must be true for the whole aggregate.

Fig. 2.10 A typical loudness function (Hellman and Zwislocki, 1961) is compared to a normalized driven firing rate of an auditory-nerve fiber (crosses; Zwislocki, 1973), integrated response of the auditory nerve (filled circles; Teas et al., 1962) and the strength of the stapedius-muscle reflex (unfilled circles; Zwislocki and Shepherd, 1972). All the characteristics tend to parallel each other. The loudness function and driven neural activity converge on a linear relationship to sound intensity at low sound intensities (intermittent straight line). Modified from Zwislocki (1974), with permission from copyright holder

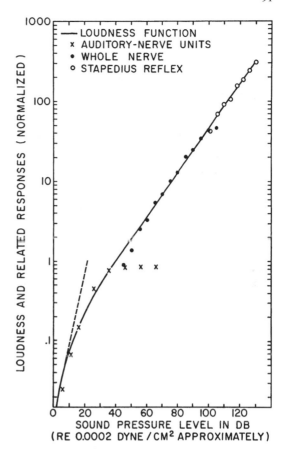

The above analysis has been applied to sensory receptors and peripheral neurons but psychological responses involve higher stages of the nervous system. Does the linearity hold there for small signals? Probably yes, for the same reasons that it holds for the periphery, but we cannot be sure. Direct empirical evidence is necessary. Several examples of such evidence have been given for the auditory system in the first section. A more explicit one is shown graphically in Fig. 2.10. The loudness curve of Fig. 2.1 (solid line) is compared to the normalized driven firing rates of a single neuron in the auditory nerve (crosses) as well as to the electrical whole-nerve response (filled circles) and the contraction strength of the stapedius muscle, as reflected in the acoustic-impedance change at the tympanic membrane (unfilled circles) (Zwislocki, 1974, 2002). Note that all the characteristics tend to parallel each other, except for the saturation section of the single-neuron characteristic. Note in particular that the single-neuron characteristic parallels the loudness curve at low signal levels. Additional evidence can be found in the next section concerning the generality of the asymptotic-linearity law.

2.3 Generality

Because of the postulated linear summation evident in Fig. 2.10, the analysis concerning single neurons can be extended to neuronal aggregates and psychophysical experiments. Auditory masking experiments in which a pure tone is masked by random noise are of particular interest. The analysis can be applied to them by simply adding the external noise intensity, N_M, to the internal noise intensity, N_I, producing the total noise intensity, $N_T = N_I + N_M$. One caveat must be observed, however. Only the portion of the extrinsic noise intensity that contributes to the masking effect should be included. Frequency bands sufficiently removed from the test-tone frequency not having such an effect must be excluded. Under these conditions, the total loudness of the tone and noise together obeys the equation (Zwislocki, 1965)

$$L_T = a(S + N_I + N_M)^\theta \tag{2.10}$$

where L_T means the total loudness, a, a dimensional constant, and θ, a power exponent on the order of 0.3 for tone frequencies that are not very low. The equation can be rewritten in the form

$$L_T = a(N_I + N_M)^\theta \left(1 + \theta S/(N_I + N_M)\right)^\theta. \tag{2.11}$$

and, for $S \ll (N_I + N_M)$, approximated by

$$L_T = a(N_I + N_M)^\theta (1 + \theta S/(N_I + N_M)) \tag{2.12}$$

For listening to the tone alone, the loudness near the threshold of detectability becomes

$$L_s = a\theta(N_I + N_M)^{\theta-1} S \tag{2.13}$$

and, according to the theory, should be directly proportional to the tone intensity. This is confirmed by the monaural loudness curves of Fig. 2.11 determined by the method of numerical magnitude balance in the absence ($N_M = 0$) and the presence of masking noise at two levels, respectively. The uppermost curve was obtained in the absence of noise, the next curve, in the presence of noise that was presented intermittently in the time gaps between tone bursts and did not produce any direct masking, for the next two curves, the noise was continuous and produced partial masking of the tone bursts. At near-threshold tone intensities, all the curves approach a linear relationship to tone intensity, as is shown by the extrapolating intermittent lines. This is in agreement with the loudness-level curves of Fig. 2.6 obtained in the same experiment and having slopes independent of the masking level near the threshold of detectability. The curves of Fig. 2.11 can be approximated over their entire range by Eq. 2.10 when the loudness of the noise is subtracted (Zwislocki, 1965).

$$L = a(S + N_I + N_M)^\theta - a(N_I + N_M)^\theta \tag{2.14}$$

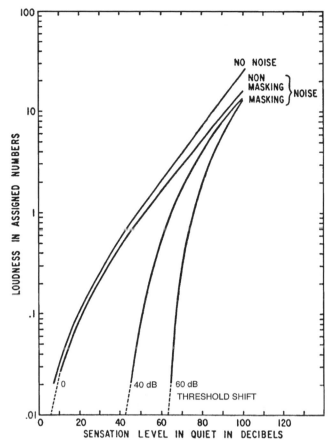

Fig. 2.11 Loudness curves at 1,000 Hz determined by numerical magnitude balance in the absence and presence of masking noise. Uppermost curve–no noise; next curve – nonmasking noise in the gaps between tone bursts; lowest two curves – masking noise at two levels presented continuously. Intermittent extrapolating lines show a linear relationship to sound intensity. Modified from Hellman and Zwislocki (1964), reproduced by permission of the American Institute of Physics

Similar subjective intensity characteristics can be obtained for the sense of touch by using vibrotactile stimuli. The similarity is only superficial, however, because the characteristics result from an entirely different mechano-receptive system, actually consisting of four systems based on four different kinds of receptors (Bolanowski et al., 1988). An example for sinusoidal vibration at a frequency of 250 Hz is shown in Fig. 2.12 with reference to sensation level (Verrillo et al., 1969). The stimulus was delivered to the thenar eminence of the right hand by a solid vibrator with a flat, circular contact surface that was pressed against the skin surface so that it produced an indentation of 0.5 mm. The surface had an area of 2.9 cm^2, and the stimuli consisted of 600 ms bursts separated by 1,400-ms time intervals. To avoid detectable transients, the stimuli were turned on and off with 10-ms ramps. The vibration

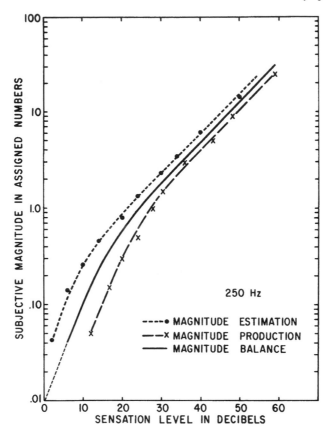

Fig. 2.12 Subjective magnitude of vibration as a function of vibration intensity. A cylindrical vibrator with a contactor area of 2.9 cm^2 was placed on the thenar eminence of the right hand and vibrated perpendicularly to the skin surface with a frequency of 250 Hz. The six observers participating in the experiment responded according to numerical magnitude balance. The resulting curve has been extrapolated downwards linearly with respect to vibration intensity. Modified from Verrillo et al. (1969), reproduced with permission from the Psychonomic Society

amplitude was measured directly with the help of an accelerometer. The subjective intensity of vibration was determined on six observers with the method of numerical magnitude balance described in Chap. 1. The filled circles of Fig. 2.12 indicate the results of magnitude estimation, the crosses, those of magnitude production. The solid line approximates their interpolated geometric means. Its extrapolation by the intermittent line obeys a linear relationship between the subjective magnitudes and the vibration amplitudes squared, proportional to intensity, in conformity with the law of asymptotic linearity.

Results obtained by the same method for several additional vibration frequencies are summarized in Fig. 2.13 by means of magnitude-balance curves. They are plotted with reference to absolute vibration amplitudes in microns rather than to sensation levels. In this way, the sensitivity differences among the various vibration

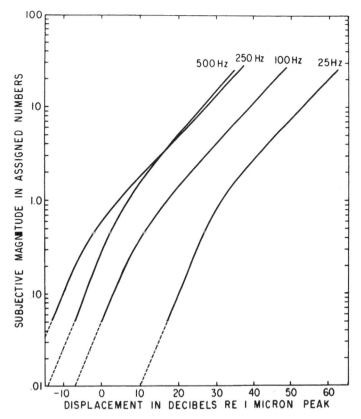

Fig. 2.13 Subjective magnitudes of vibration at four frequencies. The experimental method was the same as in Fig. 2.12. The intermittent lines extrapolate the experimental curves linearly with respect to vibration intensity. Modified from Verrillo et al. (1969), reproduced with permission from the Psychonomic Society

frequencies can be seen. Again, the extrapolating intermittent lines obey the law of asymptotic linearity with respect to vibration amplitudes squared.

The conservation of the law at all the vibration frequencies involved is of particular interest because, according to Verrillo (1966) and also Bolanowski et al. (1988), the tactile sensation at 25 Hz is mediated by different sensory receptors than that at the higher frequencies, which is mediated by Pacinian corpuscles. The preponderance of neurons ending on Pacinian corpuscles does not exhibit any spontaneous activity but the most sensitive do (Bolanowski and Zwislocki, 1984). The latter are likely the ones that determine the threshold of detectability and the shape of the subjective intensity function at near-threshold stimulus intensities. Firing-rate characteristics of one such unit are shown in Fig. 2.14 for several frequencies of vibration. Of particular interest are the curves indicating the patterns of emergence of driven firing rate from the spontaneous firing rate. They conform approximately to Eq. 2.4, thus, to the law of asymptotic linearity. A typical characteristic of a unit

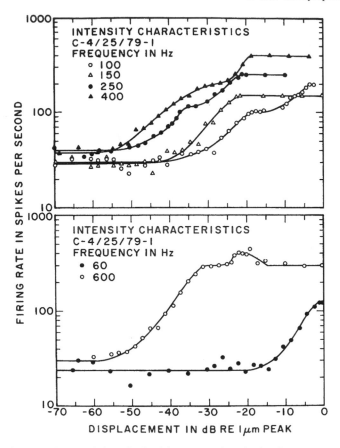

Fig. 2.14 Firing-rate characteristics of a Pacinian-corpuscle unit showing spontaneous activity. The firing rate was measured at several vibration frequencies. In the transition sections between the spontaneous and driven firing rates, the experimental curves are consistent with a linear relationship between the driven firing rate and stimulus intensity. Reproduced from Bolanowski and Zwislocki (1984), with permission from the American Physiological Society

without spontaneous activity is shown in Fig. 2.15. Its axon seems to have a high firing threshold, so that its firing pattern is not appreciably affected by the physiological noise. As a result, its action potentials tend to be synchronized with the stimulus periodicity. Such units probably control the overall response characteristics at higher vibration intensities.

Auditory and tactile receptors respond to mechanical stimuli. The input to visual receptors, the rods and cons, consists of electromagnetic waves. Does the law of asymptotic linearity still apply to vision in spite of the fundamental difference in the physical nature of the stimuli? The brightness characteristics of Figs. 2.16 and 2.17 indicate that it does (Barlow and Verrillo, 1976). The solid lines in both figures are the same and approximate the medians of absolute magnitude estimates (AMEs; Chap. 1) made by six observers at nine light intensities ranging from

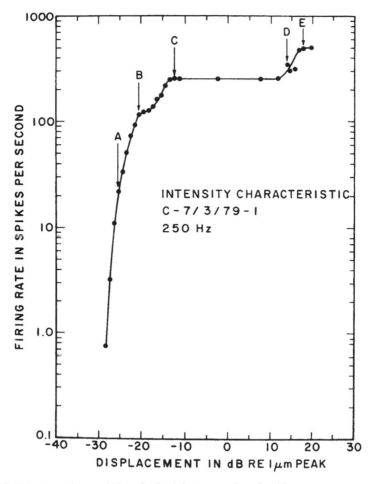

Fig. 2.15 Firing-rate characteristics of a Pacinian-corpuscle unit without spontaneous activity at 250 Hz. Reproduced from Bolanowski and Zwislocki (1984), with permission from the American Physiological Society

−9 to 0 log units (0 log ≅ 100 mLamberts) in a ganzfeld illumination. The latter was obtained by diffusing the light beam of a tungsten-filament 750 W bulb (color temperature = 2, 800°K) with half a ping-pong ball attached at the end of a blackened cone. The light intensity was controlled by means of neutral density filters. No artificial pupil was used. The observers were dark adapted before the experiment and after every light presentation that consisted of a 1-s flash. They made three brightness estimates at every intensity. The first was made for training purposes and was not included in the data averaging to minimize the response bias known to be inherent in magnitude estimation. The median brightness estimates made by the six observers are indicated in Fig. 2.16 by the crosses. The remaining symbols show the individual data obtained by taking the geometric means of two brightness estimates.

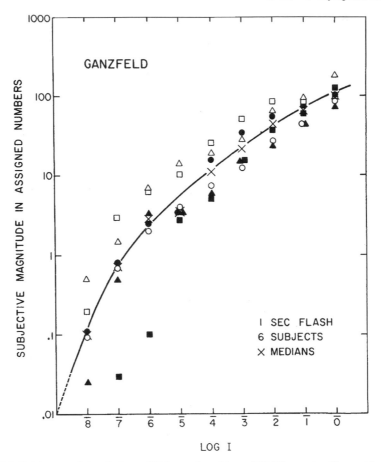

Fig. 2.16 Brightness of 1-s flashes of light produced by a 750-W tungsten-filament bulb in a ganzfeld. The brightness was measured by absolute magnitude estimation as a function of light intensity on six observers. The individual data are shown by the various symbols. The crosses indicate their medians. The solid curve is fitted to them by eye and extrapolated linearly by the intermittent line. Modified from Barlow and Verrillo (1976), with permission from Elsevier

The ganzfeld illumination was used to avoid contrast effects. To determine the effect produced by visual contrast, the light field was restricted to 2° of solid angle and presented on a black background. The median results obtained with such a field on six observers are illustrated in Fig. 2.17. In agreement with other studies, the contrast effect and the size of the illuminated field had only a small effect on brightness judgments, as indicated by the unfilled circles.

In both Figs. 2.16 and 2.17, the intermittent straight lines extrapolating the experimental data have a slope of one, indicating that the corresponding brightness functions converge on a linear relationship between the estimated brightness and light intensity. Thus, the law of asymptotic linearity applies to vision under both conditions, ganzfeld and a small target displayed on a black background.

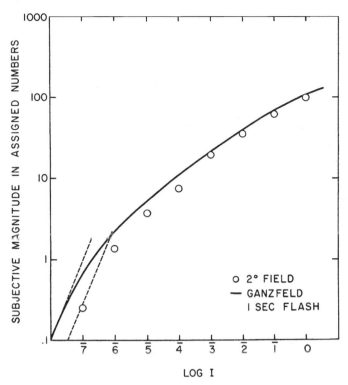

Fig. 2.17 The same as Fig. 2.16 but for a target subtending a 2° solid angle on a black background. The results, indicated by the unfilled circles, are compared to the ganzfeld results of Fig. 2.16. The intermittent lines parallel a linear function of light intensity. Modified from Barlow and Verrillo (1976), with permission from Elsevier

The results described above were obtained with dark-adapted eyes, but does asymptotic linearity hold when the eyes are light-adapted? The answer is provided by the right-hand set of data points in Fig. 2.18. They have been obtained monocularly with a circular target subtending a solid angle of 5° and displayed on a black background after the eye had been adapted for three minutes to the brightness of 106 dB re 10^{-10} Lamberts produced by illuminating a white cardboard with a flood light (J.C. Stevens and Stevens, 1963). Only the right eye was light adapted. The left eye remained dark adapted. The stimuli were presented alternately to both eyes for 2 s each, and the ten observers participating in the experiment estimated the experienced brightness magnitudes with reference to a 74-dB stimulus presented to the dark-adapted eye, to which the number 10 was assigned by the experimenter. Every stimulus was presented twice in a random order. The geometric means and the inter quartile ranges of the observers' estimates are displayed in Fig. 2.18 by means of the unfilled circles and the vertical bars, respectively. All the data points were shifted upward by a constant distance so as to bring the dark-adaptation data into coincidence with the Brill scale. They were approximated by a straight line. This

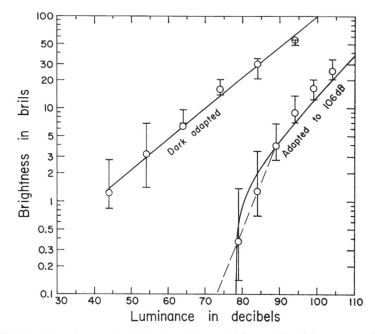

Fig. 2.18 The effect of monocular light adaptation on brightness. The right eye was adapted to 106 dB light intensity relative to 10^{-10} Lamberts (right set of data points), the left eye was dark adapted (left set of points). An illuminated target subtending a solid angle of 5° with a black surround was presented alternately to the two eyes, and ten observers estimated its brightness relative to a standard of 74 dB presented to the dark-adapted eye and called 10. The results were normalized to make the data obtained under dark adaptation coincide with the Brill scale. These data were fitted with a power function. The data obtained under light adaptation were fitted, probably erroneously, with a curve obeying Eq. 2.15. The intermittent line shows that they are linearly related to light intensity at low light intensities. The data obtained under dark adaptation do not seem to have reached sufficiently low light intensities to show the linearity effect. Modified from Stevens and Stevens (1963), with permission from copyright holder

was possible because no light intensities near the threshold of detectability were included. The data obtained for the light-adapted eyes were fitted by a line obeying Eq. 2.15

$$\psi = k(L - L_o)^\beta \tag{2.15}$$

where ψ means the brightness in assigned numbers, k, a dimensional constant, L, light intensity, L_o, the threshold light intensity, and β, an exponent on the order of 0.33. Clearly, the fit is not very good because the line misses 4 out of 6 points. On the other hand, the three lowest points fall on the intermittent line that follows direct proportionality with light intensity, in agreement with the law of asymptotic linearity.

The validity of the law of asymptotic linearity can also be tested by means of a visual experiment of an entirely different kind. It concerns visual estimation of line length. Substantial literature indicates that subjective line length is a power function of physical line length with an exponent approximating unity (e.g. Zwislocki, 1983).

A recent experiment has shown, however, that this is not true for very short, thin lines for which the exponent approaches asymptotically a value of 2 (Sanpetrino, 2005). This power exponent corresponds to the dimension of intensity and is consistent with the law of asymptotic linearity, which does refer to intensity. The experiment was not performed to test the prospective law of asymptotic linearity but, rather, to calibrate individual observers with cochlear implants for loudness measurements. The convergence of the subjective line lengths on the slope of two at very short line lengths was obtained incidentally. Thin lines, 1-mm thick, were projected horizontally on a smooth white wall of a width substantially greater than the longest line used, so that boundary effects were minimized. No marks were present on the wall from which the length of the lines could be inferred. The lines were projected with the help of a power-point set up, and their length was varied practically continuously by means of a digital controller connected to a computer. Both methods, absolute magnitude estimation and absolute magnitude production, were applied. In the first, lines of various lengths were presented 3 times in a random order; in the second, numbers were given within the range previously used by the observers. Like the lines, every number was given 3 times in a random order. The responses associated with the first set of trials for the line lengths or numbers were discarded, the responses associated with the second and third trials were pairwise geometrically averaged. Because of small differences between the magnitude-estimation and production results, both sets have been combined to form a common scatter plot shown by the circles on double-logarithmic coordinates of Fig. 2.19. To bring the trend of the data more clearly in evidence, the scatter plot has been divided into sections along the abscissa axis, and median values of the sections have been determined. They are marked by solid triangles in the figure. Finally, the medians have been approximated by two straight lines according to the least-squares method. The line extending beyond the abscissa of 3 cm has a slope of 0.95, practically 1; the line holding for smaller lengths, a slope of 2.1, practically 2. The corresponding product moment coefficients amount to 0.92 and 0.62, respectively. The modest value of the latter is due to a substantial scatter of the individual data.

Other averaging schemes were attempted, among them, polynomial curve fitting of the entire scatter plot. Although the curve slope depended on the order of the polynomial for very short lines, it remained greater than 1 for all polynomials of orders higher than 1 and hovered around 2 for polynomials between the orders of 2 and 5. In view of this finding and the evidence presented in Fig. 2.19, there can be little doubt that the law of asymptotic linearity holds for apparent length, provided the line is sufficiently thin. Thicker lines cannot be shortened enough for the phenomenon of accelerated apparent shortness to appear. They become thicker than they are long.

Available empirical data allow a test of the law of asymptotic linearity in yet another sense modality, this one concerning the sensation of warmth. The sensation is of particular interest because its threshold lies well above the null of physical heat energy. The situation is somewhat similar to that of partial auditory masking by random noise that can shift the threshold of audibility of a pure tone by a comparable amount. As shown in Fig. 1.4, the slope of the loudness function then

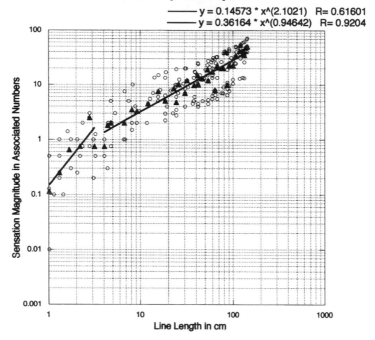

Fig. 2.19 Subjective length of thin lines measured by numerical magnitude balance on six observers as a function of the physical line length. Every observer made three estimates of every length and produced three lengths. The first response was discarded and the remaining two were averaged geometrically. The unfilled circles show the individual geometric means. Because of a small difference between the magnitude estimations and productions, the means have been combined to a common scatterplot. The scatterplot has been divided along the abscissa axis into narrow sections, and medians of the data within the sections have been determined. They are indicated by filled triangles. The medians have been fitted by two straight lines according to the least-squares method, one for the physical lengths smaller than 3 cm, the other, for greater lengths. The slope of the former obeys approximately the square of the physical length, the slope of the latter, the physical length (Sanpetrino-Anzalone, unpublished data, 2005)

becomes steeper than in the absence of the noise and the steepness increases with the threshold shift. One example of the growth of subjective warmth magnitude with heat intensity is shown in Fig. 1.20 (data from J.C. Stevens and Marks, 1971; graph modified from Marks, 1974). The stimulus consisted of a heat flux generated by a 1 kW projection lamp aimed at the forehead. The forehead was painted with India ink to facilitate heat absorption. The variable area of exposure was controlled by aluminum masks, and the exposure time, by a shutter that was opened every 30 s for 3 s. The radiant intensity was measured with a Hardy radiometer. The resulting data shown in Fig. 1.20 are based on magnitude estimates produced by 15 observers relative to reference standards they chose themselves. The heat intensity and the

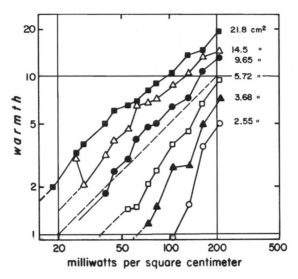

Fig. 2.20 Magnitude of warmth sensation produced by radiant heat on the forehead as a function of heat intensity and irradiated area. The heat was produced by a 1 kW projection lamp, and the forehead was painted with India ink to facilitate heat absorption. The area of exposure was controlled with aluminum masks, and the exposure duration with a shatter that was opened for 3 s every 30 s. A group of 15 observers made magnitude estimates of warmth sensation relative to standards they chose themselves. The heat intensity and the irradiated area were varied at random within the same sequence of presentations. The data points indicate geometric means of the group responses, and the points belonging to the same irradiated area are connected by straight lines. The intermittent lines are their linear extrapolations. Modified from Marks (1974)

irradiated area were varied at random in the same sequence of presentations. The data points indicate that the heat sensation grew with both heat intensity and the exposed area and grew faster with the intensity as the area was decreased. They were originally fitted by a family of smooth curves following the form of Eq. 2.15 and converging at high heat intensities on one point (J.C. Stevens and Marks, 1971). However, this fit was not very satisfactory, and other schemes were attempted, all showing one or another deficiency and running into the fundamental objection of implying a zero sensation at the threshold of detectability. This is only possible on the unrealistic assumption of the absence of internal noise. For these reasons, the data points of Fig. 2.20 have been simply joint by straight lines within each set (Marks, 1974). In addition, intermittent lines have been drawn to extrapolate the sets parallel to the slope of one indicated by the intermittent line that extends between the abscissas of 20 and 200 mW per cm². With the exception of one idiosyncratic point, the extrapolating lines seem to be consistent with the point sets and to form a curve pattern similar to that of partial auditory masking in Fig. 2.11, in agreement with the law of asymptotic linearity.

The inference that warmth sensation increases in direct proportion to heat intensity near its threshold of detectability is supported by the finding that the threshold decreases nearly in inverse proportion to exposure duration (Stevens et al., 1973).

This means that the magnitude of warmth sensation is nearly directly proportional to linear temporal summation of heat intensity.

In the preceding examples, the law of asymptotic linearity has been applied to sensory systems responding to physical stimuli, such as mechanical pressure or displacement, light or heat. Does the law apply to chemoreceptive systems of gustation and olfaction? The available experimental data suggest that it does but they are insufficient for a definitive answer. Suprathreshold responses have attracted the main attention and only a few experiments have included near-threshold stimuli. The overall results may be summarized as follows. For clearly suprathreshold stimuli, the subjective magnitudes grow according to power functions of concentrations of the chemical substances. Only a few exceptions are encountered, two prominent ones concerning sucrose (Fig. 1.39) and fructose in gustation (Moskowitz, 1970). Whereas sugars tend to produce power-function exponents for sweetness slightly greater than 1 with a mean of 1.33 (Moskowitz, 1970), acids tend to produce exponents smaller than 1 for sourness (Moskowitz, 1971). In olfaction, all the odorants seem to generate power functions with exponents smaller than unity (Cain and Moskowitz, 1974).

The dichotomy between the exponents greater and smaller than 1 is of interest for the law of asymptotic linearity. For the law to be satisfied, power functions with exponents greater than 1 must become concave upwards near the threshold of detectability, those with exponents smaller than 1, concave downwards. Unfortunately, the scarce near-threshold data reveal these trends only occasionally. They can be seen in Figs. 2.21 and 2.22, the former for the sweetness of six hexose

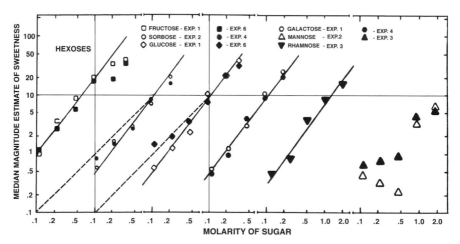

Fig. 2.21 Sweetness magnitudes of several hexose sugars as functions of concentration. The solutions involved were made of reagent-grade chemicals in Cambridge (Mass.) tap water and presented to the observers in paper cups at a temperature of 19°. The sweetness magnitudes were numerically estimated by observers selected from a group of 83 relative to standards they chose themselves. With the exception of rhamnose, the sweetness of every sugar was judged twice. The results are indicated by the various symbols and approximated by power functions (except mannose). The intermittent lines show a linear relationship to the sugar concentration. Modified from Moskowitz (1970), reproduced with permission from the Psychonomic Society

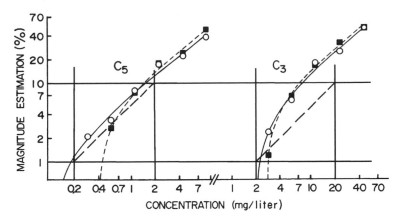

Fig. 2.22 Odorants, *n*-propanol (C₃) and *n*-pentanol (C₅) diluted in odorless air were presented to the observers through two tubes of an olfactometer at a flow rate of 4 l/min for C₃ and 6 l/min for C₅. One tube served for the adapting stimulus, the other, for the test stimulus, both containing the same odorant. The timing of the stimuli was controlled by electromechanical relays. A group of 19 observers made numerical magnitude estimates of odor intensity after adaptation to the same odor, which was achieved by taking three or eight breaths of the diluted odorant. Every observer made two estimates relative to a standard they chose themselves. Subsequently the standards were normalized to the number 10, and the individual responses were geometrically averaged. The data points show the medians of these geometric means, the unfilled circles for the three-breaths adaptation, the filled squares, for the eight-breaths one. The solid and intermittent curves were fitted to the data according to Eq. 2.15. The straight lines indicated by the long-dashes parallel the linear relationship between the subjective odor intensity and odorant concentration. Modified from Cain and Engen (1969), reproduced with permission from the Psychonomic Society

sugars (Moskowitz, 1970), the latter, for the odors of two odorants, *n*-propanol and *n*-pentanol (Cain and Engen, 1969). The data of Fig. 2.21 were obtained by the method of magnitude estimation referred to standards chosen by the observers themselves. The solutions involved were prepared of various quantities of reagent-grade chemicals and Cambridge (Massachusetts) tap water. They were presented to the observers in paper cups containing 5–10 ml of the solution to be sampled at a temperature of 19°. The different concentrations were presented in irregular order, and the observers rinsed their mouths after every presentation. Various subgroups of observers chosen from a total group of 83 participated in the experiments. The subjective sweetness of every sugar, except rhamnose, was measured in two sessions. With the exception of the results for mannose, the obtained data points followed approximately power functions. However, for small concentrations of sorbose and glucose, they partially deviated from such functions upwards, in agreement with the law of asymptotic linearity, as is evident in Fig. 2.21, where the intermittent lines indicate the slope of 1. Similar deviations occurred for some other sugars not included in the figure. That the effect was always small and did not occur in connection with all sugars was probably due to two causes – the exponents of the power functions deviated only modestly from unity, and the data were spars near the thresholds of detectability. Nevertheless, all the deviations from the power functions were consistent with the law of asymptotic linearity.

The olfactory data of Fig. 2.22 were also obtained by means of magnitude estimation relative to the number chosen by each observer for the first stimulus presented. Subsequently, all the individual data were normalized by assigning to the first stimulus the number 10, so that the data do not express absolute values. A group of 19 observers participated in the experiment, and every one of them judged the subjective intensity of every stimulus twice. The data points of Fig. 2.22 indicate the group medians of the geometric means of these pairs of judgments. The stimuli consisted of odorants diluted in odorless air in various proportions and were presented to the observers through two tubes of an olfactometer, one serving for the test odorant the other for an adapting odorant. In Fig. 2.22, both odorants consisted of the same chemicals. The flow rate amounted to 4 l/min for propanol (C_3) and to 6 l/min for pentanol (C_5). The timing of the flow was controlled by electromechanical relays. In the adapting condition, the observers were allowed to take three (unfilled circles) or eight (filled squares) breaths at a rate controlled by a metronome. In the following test condition, they took one breath. At sufficiently high concentrations, the data points followed approximately power functions with exponents smaller than unity; at lower concentrations, the slope of the curves approximated by the data points increased in apparent agreement with the law of asymptotic linearity. The data points were originally approximated, probably incorrectly, by theoretical curves obeying Eq. 2.15 and having a vertical asymptote at vanishingly small concentrations. In fact, the lowest three points for n-pentanol (C_5) obtained with weak adaptation (three breaths) are consistent with a slope of 1 indicated by the intermittent line with the longer dashes. Stronger adaptation (eight breaths) increased the slope above 1 within the experimental range of concentrations but the slope at lower concentrations remained unknown. The experiments with n-propanol (C_3) produced a similar pattern of results, except that all the slopes implied by the data points were greater. Still, the locations of the lowest two points obtained for weaker adaptation appear to be consistent with a slope of 1 (intermittent line with the long dashes) within the experimental error. Of course, the adaptation probably contributed somewhat to the steepness of the curve implied by the points.

Although the available experimental data on chemoreception do not conclusively support the law of asymptotic linearity, they are not in conflict with it.

A definitive decision will have to await further experimentation.

Multisensory examples of empirical results given above, together with its biophysical basis, suggest that the proposed law of asymptotic linearity may have universal validity in psychophysics. Of course, the universality of an empirical law can never be definitely established and must be continually tested in quest of possible exceptions.

References

Barlow, H.B. Dark and light adaptation: Psychophysics. In: *Handbook of Sensory Physiology VII/4; Visual Psychophysics*, pp. 1–28. D. Jameson and L.M. Hurvich (Eds.), Springer, New York, 1972.

Barlow, R.B., Jr., and Verrillo, R.T. Brightness sensation in a ganzfeld. *Vision Res.* 16: 1291–1297, 1976.

Bolanowski, S.J., and Zwislocki, J.J. Intensity and frequency characteristics of Pacinian corpuscles. I. Action potentials. *J. Neurophys.* 51(4): 793–811, 1984.

Bolanowski, S.J., Gescheider, G.A., Verrillo, R.T., and Checkosky, C.M. Four channels mediate the mechanical aspects of touch. *J. Acoust. Soc. Am.* 84(5): 1680–1694, 1988.

Cain, W.S., and Engen, T. Olfactory adaptation and the scaling of odor intensity. In: *Olfaction and Taste*, pp. 127–141. C. Pfaffmann (Ed.), Rockefeller University Press, New york 1969.

Cain, W.S, and Moskowitz, H.R. Psychophysical scaling of odor. In: *Human Responses to Environmental Odors*, pp. 1–31. Academic Press, New York, 1974.

Feldtkeller, R., Zwicker, E., and Port, E. Lautstärke, Verhältnislautheit und Summenlautheit. *Frequenz* 13: 108–117, 1959.

Fuortes, M.G.F. Generation of responses in receptor. In: *Handbook of Sensory Physiology*, I.W.R. Lowenstein (Ed.), Springer, Berlin-Heidelberg-New York, 1971.

Green, D.M., and Swets, J. *Signal Detection Theory and Psychophysics*. Wiley, New York, 1966.

Hellman, R.P., and Zwislocki, J.J. Some factors affecting the estimation of loudness. *J. Acoust. Soc. Am.* 33. 607–694, 1961.

Hellman, R.P., and Zwislocki, J.J. Monaural loudness function at 1000 cps and interaural summation. *J. Acoust. Soc. Am.* 35: 856–865, 1963.

Hellman, R.P., and Zwislocki, J.J. Loudness sensation of a 1000-cps tone in the presence of a masking noise. *J. Acoust. Soc. Am.* 36: 1618–1627, 1964.

Hellman, R.P., and Zwislocki, J.J. Loudness determination at low sound frequencies. *J. Acoust. Soc. Am.* 43(1): 60–64, 1968.

Kiang, N.Y.-S. A survey of recent developments in the study of auditory physiology. *Trans. Amer. Otol. Soc.* 66: 108–116, 1968.

Marks, L.E. *Sensory Processes: The New Psychophysics*. New York: Academic Press, 1974.

Miller, G.A. Sensitivity to changes in the intensity of white noise and its relation to masking and loudness. *J. Acoust. Soc. Amer.* 19: 609–619, 1947.

Moore, B.C.J. Testing the concept of softness imperception: Loudness near threshold for hearing-impaired ears. *J. Acoust. Soc. Am.* 115(6): 3103–3111, 2004.

Moskowitz, H.R. Ratio scales of sugar sweetness. *Percept. Psychophys.* 7(5): 315–320, 1970.

Moskowitz, H.R. Ratio scales of acid sourness. *Percept. Psychophys.* 9(3B): 371–374, 1971.

Robinson, D.W. The subjective loudness scale. *Acustica* 7: 217–233, 1957.

Sanpetrino, N.M. Unpublished data, 2005.

Scharf, B. Loudness. In: *Handbook of Perception: Vol. 4. Hearing*, pp. 187–242. E.C. Carterette and M.P. Friedman (Eds.), Academic, New York, 1978.

Scharf, B., and Stevens, J.C. The form of the loudness function near threshold. *Proceedings of the 3rd International Congress on Acoustics*, Elsevier, Amsterdam, pp. 80–82, 1961.

Stevens, S.S. The direct estimation of sensory magnitudes-loudness. *Am. J. Psychol.* 69: 1–25, 1956.

Stevens, J.C., and Marks, L.E. Spatial summation and the dynamics of warmth sensation. *Percept. Psychophys.* 9(5): 391–398, 1971.

Stevens, J.C., and Stevens, S.S. Brightness function: Effects of adaptation. *J. Opt. Soc. Am.* 53: 375–385, 1963.

Stevens, J.C., Okulicz, W.C., and Marks, L.E. Temporal summation at the warmth threshold. *Percept. Psychophys.* 14(2): 307–312, 1973.

Teas, D.C., Eldredge, D.H., and Davis, H. Cochlear responses to acoustic transients: An interpretation of whole-nerve action potentials. *J. Acoust. Soc. Am.* 34: 1438–1459, 1962.

Verrillo, R.T. Effects of spatial parameters on the vibrotactile threshold. *J. Exp. Psychol.* 71: 570–574, 1966.

Verrillo, R.T., Fraioli, A., and Smith, R.L. Sensory magnitude of vibrotactile stimuli. *Percept. Psychophys.* 6(A): 366–372, 1969.

Zwicker, E., and Feldtkeller, R. *Das Ohr als Nachrichtenempfanger*. S. Hirzel Verlag, Stuttgart, 1956.

Zwislocki, J.J. Analysis of some auditory characteristics. In: *Handbook of Mathematical Psychology*, Vol. 3, pp. 1–97. R.D. Luce, R.R. Bush, and E. Galanter (Eds.), Wiley, New York, 1965.

Zwislocki, J.J. On intensity characteristics of sensory receptors: A generalized function. *Kybernetik* 12: 169–183, 1973.

Zwislocki, J.J. A power function for sensory receptors. In: *Sensation and Measurements*. H.R. Moskowitz, B. Scharf, and J.C. Stevens (Eds.), Reidel, Dordrecht, Holland, 1974.

Zwislocki, J.J. Group and individual relations between sensation magnitudes and their numerical estimates. *Percept. Psychophys.* 33: 460–468, 1983.

Zwislocki, J.J. Auditory system: Peripheral nonlinearity and central additivity, as revealed in the human stapedius-muscle reflex. *Proc. Nat. Acad. Sci. USA* 99(22): 14601–14606, 2002.

Zwislocki, J.J., and Shepherd, D.C. Unpublished data, 1972.

Chapter 3
Law of Additivity

3.1 Definition and Consistency with Ratio Scaling

According to the Law of Additivity, sensation magnitudes and other subjective magnitudes sum linearly when processed independently up to the stage of summation. The law may be regarded as established for most same-modality sensation magnitudes but not heteromodality ones. Because it may be applied to the generation of functions that relate subjective magnitudes to their underlying stimulus magnitudes, it is often used for proving the Power Law, as is done in Chap. 1. The procedure consists preferentially of presenting two subjectively equal magnitudes and assigning to their sum a numeral that indicates magnitude doubling. For example, if the numeral 1 is assigned to the component magnitudes, the numeral 2 would be assigned to their sum.

The concept of additivity of subjective magnitudes was introduced at least as early as 1920s by communication engineers for the purpose of determining the growth of loudness as a function of tone intensity. Finding that loudness growth with sound intensity was not well predicted by Fechner's Law, Fletcher and his associates at Bell-Telephone Laboratories searched for empirical methods that could determine it (Fletcher, 1929). Intuition told them that a tone delivered to both ears should sound twice as loud as a tone delivered to one ear alone. Their intuition seemed to be confirmed by subjective impression – a tone presented binaurally did seem to sound twice as loud as a monaural tone at the same physical intensity (Fletcher and Munson, 1933). As a further step, they increased the intensity of the monaural tone to make its loudness match the loudness of the binaural one. In this way, they were able to find the intensity ratio corresponding to doubling of loudness. By continuing the procedure, they were able to construct the function relating loudness to sound intensity. To validate the procedure, Fletcher (1935) presented simultaneously two tones of equal loudness but different frequencies to the same ear and compared the resulting loudness to the loudness of the single tone. As long as the frequency separation was sufficient to prevent a destructive interaction, the two tones seemed to be twice as loud as a single tone. By matching consecutively the loudness of the component tones and that of their sum to a third tone, Fletcher found roughly the same relationship between loudness and sound intensity as in the binaural experiment. Fletcher and Munson extended the monaural procedure to multiple tones and further confirmed the relationship between the two variables obtained initially.

In the same time period, other investigators used more direct procedures to obtain loudness scales. Observers were requested to adjust the loudness of one tone to make it twice or half as loud as that of another tone. Other loudness ratios were used as well. In some experiments, the observers were required to judge the loudness ratios between tones of preset intensity. According to Stevens and Davis's (1938) review, through their agreement, these more direct procedures, although less precise, validated the procedures based on summation. The relationships for half loudness and 0.1 loudness are shown graphically in Fig. 3.1. They are expressed in terms of intensity levels producing loudness equality. The agreement among the various experiments based on direct fractionation and between them and summation experiments is rather impressive, especially for loudness halving. It may be regarded as an early validation of the Law of Additivity.

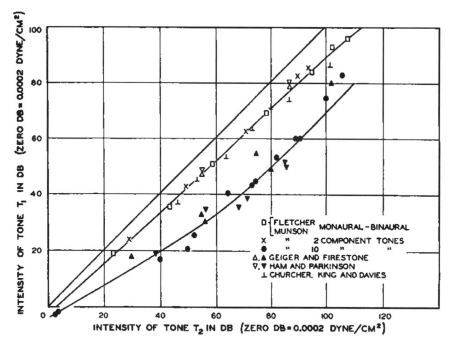

Fig. 3.1 The ordinate shows the SPL of a tone that appears half as loud as the same tone at a SPL indicated by the abscissa (unfilled triangles and inverted T symbols) or one tenth as loud (filled triangles). The unfilled squares show the SPLs of a tone presented binaurally (ordinate) and monaurally (abscissa) for loudness equality. The slanted crosses and filled circle show the SPLs of two-component, respectively, ten-component sounds (ordinate) that sound equally loud as one-component sounds. Reproduced from Stevens and Davis (1938), with permission from the American Institute of Physics

3.2 Further Validation of the Law for Loudness

Loudness appears to have been the preferred modality for experiments on the additivity of subjective magnitudes. Many concerned the loudness of binaural tones, as exemplified by the early experiments described in the preceding section. Perhaps the most fundamental ones were dedicated to proving that the loudness magnitude of a tone heard binaurally is, indeed, equal to twice the loudness magnitude of the same tone heard monaurally; stated more generally, that the monaural loudness magnitudes are interaurally additive.

One experimental series utilized a binaural loudness function established by several methods that produced mutually consistent results (Hellman and Zwislocki, 1963). The results, indicated by the various symbols in Fig. 3.2, refer to loudness ratios rather than to absolute loudness magnitudes. Nevertheless, they are anchored

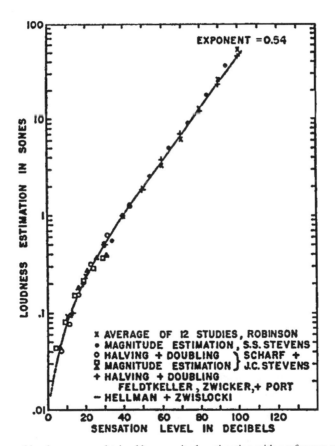

Fig. 3.2 Binaural loudness curve obtained by magnitude estimation with a reference standard consistent with AME. The various symbols refer to loudness ratios determined by a number of investigators using various methods. Reproduced from Hellman and Zwislocki (1963), with permission from the American Institute of Physics

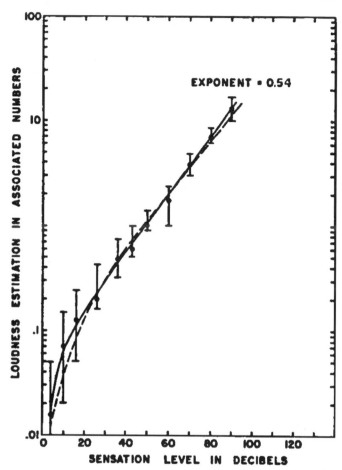

Fig. 3.3 The solid line indicates the monaural loudness curve obtained on nine observers by AME. The solid circles and vertical lines refer to geometric means and standard deviations of the group. The intermittent curve has been reproduced from Fig. 3.2. Reproduced from Hellman and Zwislocki (1963), with permission from the American Institute of Physics

on the solid curve based on reference standards consistent with such magnitudes (Hellman and Zwislocki, 1961; also Chap. 1). Corresponding monaural loudness curves were determined by two methods. One similar to the method with the reference standards, used in the binaural experiments, the other based on absolute magnitude estimations without a reference standard (Hellman and Zwislocki, 1963). Both curves are shown in Fig. 3.3, the first by means of the intermittent line, the second by means of the solid one. The group means and standard deviations for the latter are also given. Both the binaural curve of Fig. 3.2 and the monaural curves of Fig. 3.3 are reproduced in Fig. 3.4. To the binaural curve obtained with designated standards and indicated by the intermittent line is added a theoretical curve corresponding to the monaural curve obtained without designated standards. This is done

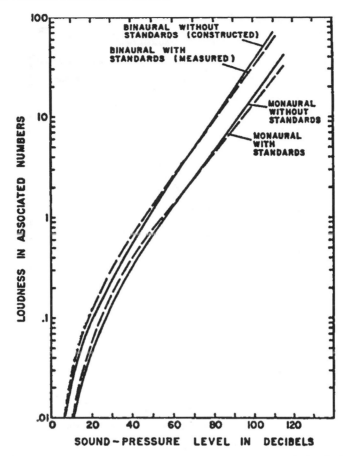

Fig. 3.4 Solid lines refer to binaural and monaural loudness curves, respectively, obtained by magnitude balance and derived from the curves of Figs. 3.2 and 3.3. Intermittent lines correspond to loudness curves determined with reference standards. Reproduced from Hellman and Zwislocki (1963), with permission from the American Institute of Physics

on the assumption that designated standards have the same effect in terms of loudness ratios on the binaural curve as on the monaural one. The reader should note that the loudness ratio between the corresponding binaural and monaural curves remains roughly the same over the whole extent of the curves, even where the curves become steeper near the threshold of detectability. It amounts approximately to loudness doubling and indicates a linear process of summation.

A more detailed analysis of binaural loudness summation indicates, however, that there are slight sensitivity differences between the two ears, so that the loudness of a binaural sound should not be expected to be exactly twice the loudness experienced in anyone of the two ears. Rather, it should be twice the loudness averaged between them. According to Shaw et al. (1947) the average interaural threshold difference amounts to 3.8 dB, roughly, 4 dB. According to Hellman and Zwislocki (1961), the

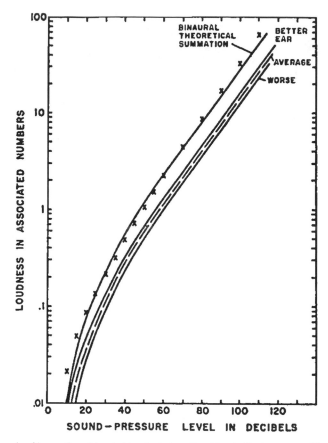

Fig. 3.5 The pair of lower lines bisected by the intermittent line indicate monaural loudness curves spaced by the average interaural sensitivity difference; the upper solid line indicates their theoretical summation. The crosses indicate the empirical binaural loudness values. Reproduced from Hellman and Zwislocki (1963), with permission from the American Institute of Physics

difference is preserved at suprathreshold levels. The relationship is illustrated in Fig. 3.5. The monaural loudness curves, spaced by 4 dB along the sound-pressure axis are shown by the thin curves; the intermittent line shows their SPL average. The ordinates of the thick curve are equal to the sums of the ordinates of the monaural curves. The crosses indicate the empirical loudness magnitudes derived from Fig. 3.2 through appropriate conversion of SLs to SPLs. The agreement between the numerical values obtained by the theoretical summation and the empirical values demonstrates binaural additivity of monaural loudness magnitudes.

The additivity was further confirmed by comparing the results shown in Fig. 3.5 to the results obtained in the past by Fletcher and Munson (1933) who determined pairs of SPLs producing loudness equality between monaurally and binaurally presented tones. To obtain corresponding SPLs from Fig. 3.5, horizontal cuts were made through the family of curves of the figure. The average SPLs so determined

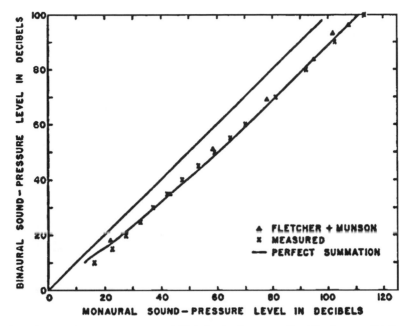

Fig. 3.6 Binaural SPL versus monaural SPL for loudness equality. The lower solid line was obtained by sectioning horizontally the mean monaural and binaural curves of Fig. 3.5; the values marked by slanted crosses were derived in an analogous fashion from the empirical values marked by slanted crosses in the same figure; the values marked by filled triangles were obtained empirically by Fletcher and Munson in 1933. The upper solid line marks perfect interaural symmetry. Reproduced from Hellman and Zwislocki (1963), with permission from the American Institute of Physics

for the monaural tones were plotted as abscissas of the solid curve in Fig. 3.6, the SPLs of the equally loud binaural tones, as its ordinates. The empirical pairs of SPL were derived from Fig. 3.5 in a similar fashion and are marked by slanted crosses. The numerical values of Fletcher and Munson are marked by filled triangles. They lie close to the values obtained by Hellman and Zwislocki (1963).

Perhaps the most direct confirmation of binaural additivity of monaural tones was obtained by Marks (1978) who measured the loudness of binaural and monaural tones on the same observers by the method of magnitude estimation with reference standards chosen by the observers themselves. As already mentioned, the method is closely related to the AME method. The measurements were performed at three sound frequencies – 100, 400 and 1,000 Hz, and involved interaurally unequal loudness magnitudes in addition to the equal ones. Specifically, the binaural loudness was measured as a function of SPL in one ear while the SPL in the contralateral ear was kept constant at several levels. The procedure was expected to produce loudness increments independent of the total loudness in the presence of binaural loudness additivity. Graphically, it was expected to generate families of parallel loudness curves spaced by loudness steps corresponding to the SPL steps. Examples of such

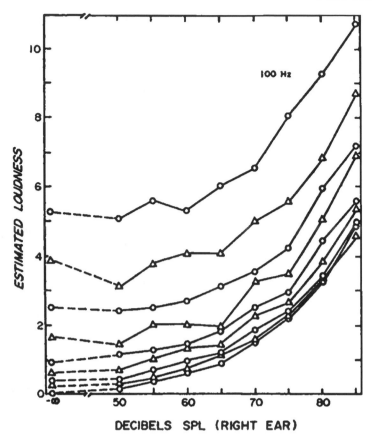

Fig. 3.7 Magnitude-estimation curves of binaural loudness summation of interaurally unequally loud tones of 100 Hz. The curves go through geometric means of group values and are plotted on a linear ordinate scale over the abscissa scale in decibels, referring to the right ear. The SPL difference between the two ears was varied parametrically. Reproduced from Marks (1978), with permission from the American Institute of Physics

families obtained on partially overlapping groups of 14 observers each are shown in Figs. 3.7 and 3.8, the former for the sound frequency of 100 Hz, the latter for that of 1,000 Hz. The curves, which follow the geometric means of the individual loudness estimates, are roughly parallel, as they should be in the presence of loudness additivity. They follow approximately power functions and would appear as straight lines on double-logarithmic coordinates. This is shown in Fig. 3.9 for a subset of the data limited to equally loud sounds in both ears. The circles and squares refer to monaural loudness magnitudes in the left and right ears, respectively, the triangles, to the corresponding binaural magnitudes. All 3 sound frequencies – 100, 400 and 1,000 Hz, are included, and the functions have power exponents of approximately 0.6 for the higher frequencies and 0.75, for the lowest, in rough agreement with the results obtained previously by absolute magnitude-balance (e.g. Hellman and Zwislocki,

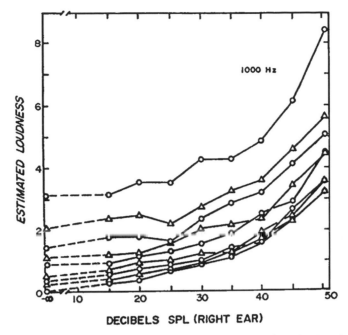

Fig. 3.8 Same as Fig. 3.7 for 1,000 Hz. Reproduced with permission from the American Institute of Physics

1968). Note that, in agreement with the latter, the 100-Hz curves are slightly concave downwards. The ordinates of the binaural curves are approximately twice the ordinates of the monaural curves for all three frequencies, indicating loudness doubling in conformity with the earlier results of Hellman and Zwislocki (1963).

Both studies, that of Hellman and Zwislocki and that of Marks indicate binaural loudness additivity. Such additivity was already assumed by Fletcher and Munson (1933) for the purpose of generating loudness functions and was validated by the agreement of their results with those produced by magnitude balance, as shown in Fig. 3.6 (Hellman and Zwislocki, 1963), and magnitude estimation (Marks, 1978). Binaural loudness additivity was also found by Levelt et al. (1972) who compared the loudness in the two ears by the method of paired comparisons and analyzed their results with the help of the theory of conjoint measurement (Luce and Tukey, 1964).

Binaural loudness additivity consistent with loudness growth according to a power function having an exponent of about 0.6 was not found by all experimenters. Caussé and Chavasse (1942) found a difference of only 6 instead of 10 dB for loudness equality between binaural and monaural tones at medium SLs. Interestingly, they found a difference of 3 dB at low SLs, which is consistent with loudness additivity in the presence of a linear loudness growth (e.g. Hellman and Zwislocki, 1963, 1968; see Chap. 2). The less than perfect summation at higher SLs may have been due to asymmetrical loudness matching in which only the monaural tone intensity was varied (Hellman and Zwislocki, 1963). An incomplete summation seems to

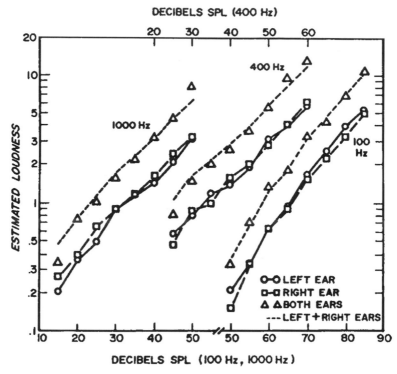

Fig. 3.9 Data of Figs. 3.7 and 3.8 with added data for 400 Hz are plotted on log. ordinate scale for interaural loudness equality. Reproduced from Marks (1978), with permission from the American Institute of Physics

have been obtained also by Scharf and Fishkin (1970) in spite of the fact that they used magnitude estimation and production procedures. These procedures are not always bias free, however, unless precautions mentioned in Chap. 1 are observed, especially, when relative rather than absolute loudness estimates are made. According to Sharf and Fishkin, binaural loudness was equal on the average to 1.7 the monaural loudness rather than 2. The less than perfect apparent summation may have been due to an artifactually reduced slope of their loudness function, which obeyed an exponent of 0.5 rather than 0.6. When the exponent is corrected by multiplication to 0.6, a perfect loudness summation is obtained (Marks, 1978). In general, lack of perfect binaural loudness summation seems to have occurred in studies in which an experimental bias of one sort or other could be demonstrated. Studies consistent with the best established course of the loudness functions appear to indicate binaural additivity.

Marks et al. (1991) were able to show that binaural summation remains unaffected when the tones in the two ears are not at exactly the same sound frequency. They presented simultaneous dichotic pairs of 1-s tone bursts at the frequency of 1,000 Hz in one ear and of 1,000, 1,040, and 1,080 Hz in the other, as well as

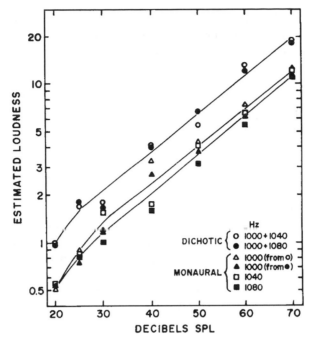

Fig. 3.10 Binaural loudness summation of equally loud simultaneous tones at slightly different sound frequencies, determined by magnitude estimation. The circles refer to binaural tone pairs, the remaining symbols, to the monaural tones. Reproduced from Marks et al. (1991), with permission from the Psychonomic Society

monotic tone bursts at the same frequencies. The tones were presented at 7 SPLs between 20 and 70 dB. The 16 observers participating in the experiments judged the loudness magnitudes of the tone bursts by the method of magnitude estimation relative to reference standards chosen by themselves. The experiments were counterbalanced by rotating the earphones between the two ears. The group results are displayed in Fig. 3.10 on logarithmic coordinates. The geometric means of the magnitude estimates are indicated by the various symbols and interpolated by solid lines. One line was sufficient for all the dichotic results, two parallel once were needed for the monotic results referring to different frequency combinations. Above an SPL of 30 dB, all the lines are straight indicating loudness growth according to power functions. Along the SPL axis the space between the dichotic and monotic curves amounts to about 10 dB, which would mean loudness doubling if the slope of the lines conformed with a power exponent of 0.6, expected for unbiased loudness functions. In fact, the curves follow an exponent of only 0.5, probably due to a bias caused by the lack of experience of the majority of the observers with loudness scaling. The exponent of 0.5 in combination with a horizontal spacing of 10 dB is consistent with a binaural/monaural loudness ratio of 1.7, similar to that obtained by Scharf and Fishken. When the exponent is corrected to 0.6, in agreement with the exponent prevailing for loudness functions that result when great care is taken to avoid biases, perfect binaural loudness summation is obtained.

Perhaps the greatest significance of the experiments of Marks, Algom and Benoit lies in the finding that perfect (after exponent correction) binaural loudness summation is maintained even for tones that are not identical in sound frequency and, therefore, in pitch. Thus, loudness and pitch must constitute separate, noninteracting attributes of sound. Experiments described below further strengthen this notion.

Already Fletcher (1935) used pairs of simultaneous, monaural tones of different sound frequencies to study the summation of their loudness magnitudes. He found that the summation was equivalent to binaural summation of tones of the same sound frequency as long as the frequency separation between them was sufficient. His results were later confirmed by those of Zwicker et al. (1957) who used multiple tone complexes. The loudness of the complexes increased with the frequency separation of the components, presumably because of decreasing mutual interaction.

As variants of Fletcher's procedure, experiments were performed in which the component tones were not presented simultaneously but separated by various time intervals. In one set of experiments, triads of 10-msec tone bursts having different sound frequencies were presented monaurally to seven observers (Zwislocki et al., 1974). The time interval between the first two bursts was variable, that between the second and third bursts was kept constant at 500 msec. The observers had to adjust the loudness of the third burst to match the combined loudness of the first two bursts or to the loudness of the second burst alone. The purpose of the latter, auxiliary, procedure was to determine the effect of the first burst on the loudness of the second. In the first and second sequences, the first burst was at a sound frequency of 1 kHz and the second at that of 4 kHz. The third burst was at 2 kHz, the geometric mean of 1 and 4 kHz, in the first sequence and at 4 kHz in the second and third ones. In the third sequence, all the bursts were at the same sound frequency of 4 kHz. As shown in Fig. 3.11, independent of sound frequency, the first burst had only a negligible effect on the loudness of the second. The combined loudness magnitude of the first

Fig. 3.11 Loudness summation of two tone bursts at two different frequencies (1,000 and 4,000 Hz) separated by a time interval of 500 msec. A third burst at 4,000 Hz or 2,000 Hz was compared in loudness either to the tone pair or to the second burst in the pair. The data points and interpolating curves refer to corresponding loudness levels. Reproduced from Zwislocki et al. (1974), with permission from the Psychonomic Society

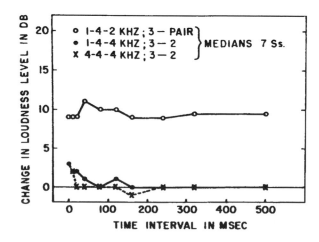

and second bursts having the sound frequencies of 1 and 4 kHz, respectively, was equal to the linear sum of their loudness magnitudes, as judged from the loudness-level difference of approximately 10 dB. The loudness integration spanned a time interval of at least 500 msec. Apparently, the loudness of the second burst was added linearly to the memorized loudness of the first.

In another set of experiments, described in Chap. 1 (Zwislocki, 1983), 20-msec tone bursts separated by 50-msec time intervals and having sound frequencies of 1 and 4 kHz, respectively, were presented monaurally with a repetition rate of $1 \sec^{-1}$. The 1-kHz bursts were presented at a number of SL, and the SLs of the 4-kHz bursts were adjusted so as to make the bursts, respectively, 0.5, 1.0, and 2.0 times as loud as the 1-kHz bursts. The loudness magnitudes of the component bursts and of the burst pairs were determined by the method of absolute magnitude balance. As described in Chap. 1, the loudness of the burst pairs proved to be approximately equal to the linear sum of the loudness magnitudes of the single bursts.

The two sets of experiments confirmed Fletcher's assumption that the loudness magnitudes of tone bursts of sufficiently different sound frequencies are additive. As an extension of this finding, they show that the process of addition can be extended over a considerable span of time. Together with the demonstration of binaural (diotic) and dichotic loudness additivity, the experiments demonstrate that the law of additivity of subjective magnitudes holds for loudness.

3.3 Generality

Experiments performed in other sense modalities indicate that additivity of sensation magnitudes is not limited to loudness but is a more general sensory phenomenon. However, it occurs only under specific conditions that may differ among the senses. Among the non-chemical senses, the conditions encountered in vibrotaction appear to be the most similar to those in hearing, those encountered in vision, the most dissimilar and the most restricted.

In the sense of touch, probably the most extensive quantitative experiments were performed on the glabrous skin of the hand by means of vibrators. They made it possible to study response characteristics paralleling those of hearing. For example, the threshold of vibration detectability was measured as a function of vibration frequency. On the basis of anatomy and, by varying the size of the vibrating contactor, Verrillo and his associates (e.g. Verrillo, 1968) were able to establish that the most sensitive vibration receptors were the Pacinian corpuscles, especially in the midfrequency range, around 250 Hz. Their sensitivity decreased toward both high and low frequencies. At low frequencies, other receptors, summarily called "Non-Pacinian" determined the vibrotactile threshold. The relationships are schematized in Fig. 3.12 according to the analysis of Bolanowski and his associates (e.g. Bolanowski et al., 1993). The finding signifies that, under appropriate conditions, the vibrotactile information is conveyed by Pacinian corpuscles at medium frequencies and by the

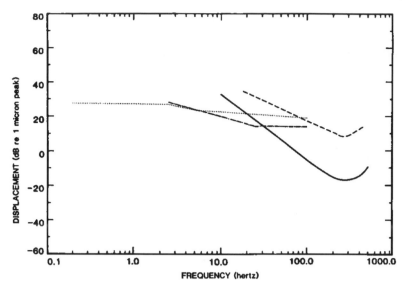

Fig. 3.12 Thresholds of detectability of Pacinian (continuous line) and three None Pacinian vibrotactile receptors. Reproduced from Bolanowski et al. (1993), with permission from Taylor & Francis Group, LLC

Non-Pacinian ones at low frequencies, indicating that the two processes are separated before the stage of their integration.

Exploiting the separation, Verrillo and Gesheider (1975) performed experiments on 5–6 observers, analogous to the auditory experiments of Zwislocki et al. (1974), in which two short bursts of sinusoids of different frequencies were followed by a third burst of intermediate frequency. The time interval between the first two bursts was varied between 35 and 500 msec, and the third burst followed the second burst at a time interval of 700 msec. The first burst was at a frequency of 300 Hz, the second at that of 25 Hz, and the third at that of 80 Hz. The observers had to set the first two bursts at equal subjective intensity, and adjust the third burst to match the total subjective magnitude of the first two bursts combined. The result is shown in Fig. 3.13 by the unfilled circles and the interpolating solid line. Unlike in hearing, the adjusted magnitude depends on the time interval between the first two bursts. At the shortest interval, it is equal to the arithmetic sum of the individual magnitudes of the two bursts, as indicated by an increment of 6 dB. Because the subjective vibrotactile magnitudes in the range of the experiments grew with vibration intensity according to a power function with an exponent of about 0.5, the 6-dB ratio amounts to magnitude doubling, indicating additivity. The magnitude decrement occurring with the increasing time interval between the stimulus bursts suggests an effect of decreasing memory.

For comparison purposes, the intermittent line in the figure shows what happens when the stimulus bursts are at similar vibration frequencies so that they are processed in the same Pacinian system before being integrated. The integration

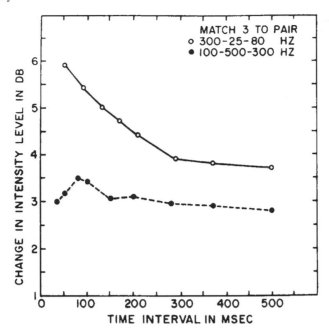

Fig. 3.13 Intensity level of a vibration burst at 80 Hz matched in subjective intensity to a preceding burst pair at 300 and 25 Hz respectively (unfilled circles) and the intensity level of a vibration burst at 300 Hz matched in subjective intensity to a preceding burst pair at 100 and 500 Hz respectively (filled circles). The abscissa scale refers to the inter-burst time interval in the pairs. Reproduced from Verrillo and Gesheider (1975), with permission from the Psychonomic Society

produces a magnitude increment of only 3 dB, corresponding approximately to the summation of stimulus energies rather than to the summation of the subjective magnitudes. This is not a trivial outcome because the bursts were not presented simultaneously so that their energies could not have been summated directly in the stimulus domain.

The subjective magnitude additivity was confirmed by Marks (1979) who used roughly the same experimental paradigm as he did for audition (Marks, 1978). A group of six subjects was involved. The stimuli consisted of 1-sec bursts of 250- and 20-Hz vibration, respectively, presented through a contactor with a contact area of 0.64 cm^2 singly and in combination. In one experiment, their intensities were arranged according to a 49 sensation-level matrix so that every stimulus component was presented at 7 different levels. Every observer judged the subjective magnitudes of the 2-frequency stimuli 6 times, and the results were expressed in terms of geometric means. The geometric means of the group results are shown in Fig. 3.14 as functions of SLs of the 250-Hz bursts. Every curve belongs to a different added 20-Hz tone burst with its SL marked at the curve. The parallel course of the curves indicates that the increments were added arithmetically.

In a supplemental experiment, the 250- and 20-Hz stimuli were presented singly as well as in matched pairs at 7 suprathreshold SLs. Their subjective intensities

Fig. 3.14 Subjective intensity of pairs of simultaneous vibration bursts at 250 and 20 Hz, respectively, plotted on a linear scale over the sensation level of the 250-Hz burst. (From Marks, 1979, with permission granted from copyright holder)

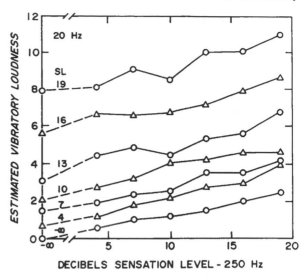

were determined by magnitude estimation without designated reference standards. A group of four observers judged every stimulus four times. The geometric means of the group responses are shown in Fig. 3.15 by various symbols as functions of the respective stimulus levels. The magnitudes of the responses to the combined stimuli are plotted by triangles over the 250-Hz axis. For comparison, the upper intermittent line follows the arithmetic sums of the corresponding subjective magnitudes of the 250- and 20-Hz stimuli. The agreement between the empirical responses to the combined stimuli and the theoretical sums indicates that the observers summed the subjective magnitudes of the component stimuli linearly.

The results of the experiments of Verrillo and Gescheider and of Marks clearly show that the law of additivity applies to vibrotaction. Additional experiments would be required to see if it applies to the sense of touch more generally.

The visual system tends to average the brightness of visual stimuli rather than summating it. Thus far, brightness additivity was demonstrated in only one set of conditions, when both eyes were illuminated uniformly – the so-called "Ganz-feld" illumination, which is contourless. Bolanowski (1987), who discovered the phenomenon investigated it for binocular brightness equality. He found simply that in binocular illumination the light appeared twice as bright as in monocular one. Because, in monocular illumination, which he needed for comparison with the binocular one, the brightness of the light tends to fade out with time, he used short (1-s) light flashes, as did Barlow and Verrillo (1976) for the same reason when determining the Ganz-feld brightness function.

Like Barlow and Verrillo, Bolanowski produced a separate ganzfeld for each eye by back-illuminating half a ping-pong ball, trimmed carefully to fit snugly around the eye ball. He tested the fit by measuring the time required for the light to fade out when only one eye was illuminated. The uniformity of the illuminated field was achieved by mounting each half-ball at the end of a nonreflecting white

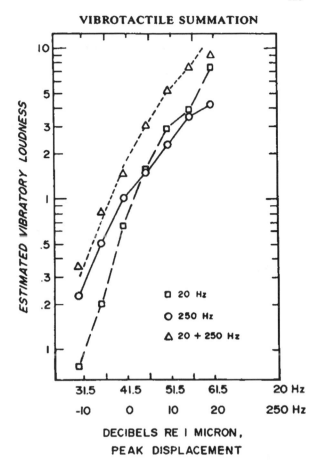

Fig. 3.15 Subjective intensity of single vibration bursts at 20 and 250 Hz and of pairs of the bursts, plotted on a logarithmic scale over peak displacement amplitudes of the vibrations. (From Marks, 1979, with permission granted from copyright holder)

cone. He used achromatic light produced by a halogen light source (Xenophot HLX64625) with appropriate optics. The light intensity was controlled by neutral-density filters and calibrated with an IL 700 Research Radiometer incorporating a CIE standard-observer curve correction. (0.0 Log I is equivalent to $3232 \, cd/m^2$ in the figures). A group of 16 naive observers participated in the experiment. After appropriate training with estimating the lengths of lines, they had to estimate the light brightness according to the method of absolute magnitude estimation. They were dark-adapted for 20 min, and their thresholds were measured before the magnitude-estimation experiments. In the latter, light intensities were presented in a quasi-random order, except that the highest intensity was not presented before the lowest one to minimize light adaptation. A group of eight observers first received a monocular set of ganzfeld stimuli in the right eye, then, a corresponding set of binocular stimuli. For another group of eight observers, this order was reversed. Some observers received a monocular stimulus in the left eye as well to check on the interocular symmetry. Because the symmetry requirement was satisfied, only the right eye was used for monocular stimuli in the main experiment. Individual brightness

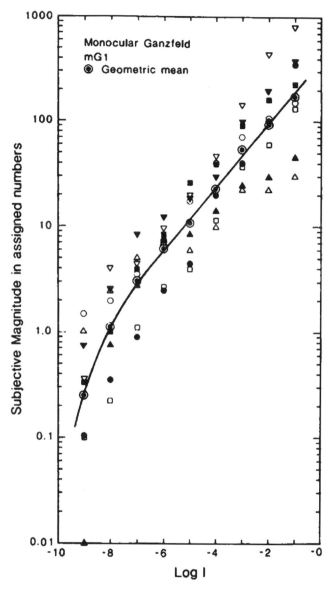

Fig. 3.16 Monocular brightness of achromatic ganzfeld light flashes determined by AME as a function of light intensity – individual data and geometric means. The curve has been fitted to the group means by eye. The abscissa scale is logarithmic and referred to $3,232 \, \text{cd/m}^2$. (Reprinted from Bolanowski, 1987, with permission from Elsevier)

estimates produced by the first group of observers, as well as their geometric means, are shown in Fig. 3.16 as functions of light intensity. The geometric means are interpolated by eye with the result indicated by the means of the solid curve. Above about -8 log units, the latter follows a power function with an exponent of about 0.29,

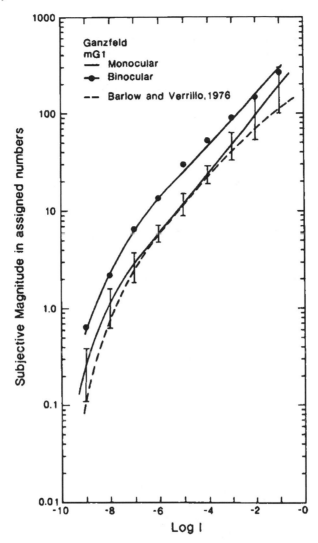

Fig. 3.17 Binocular brightness of achromatic ganzfeld light flashes determined by AME as a function of light intensity referred to $3,232\,\mathrm{cd/m^2}$. Filled circles indicate the geometric group means; the solid line has been fitted to them by eye. The thin line indicates the monocular brightness reproduced from Fig. 3.16, and the vertical lines, the corresponding standard deviations of the individual data. For comparison, the intermittent line indicates analogous data obtained by Barlow and Verrillo (1976). (Reprinted from Bolanowski, 1987, with permission from Elsevier)

in rough agreement with the comparable results of Barlow and Verrillo (1976) shown in the first chapter. The monocular curve of Fig. 3.16 together with the standard-error bars of the individual brightness estimates is reproduced in Fig. 3.17 for comparison with the results of the binocular experiment shown in terms of the geometric means and an interpolating curve. The intermittent line indicates the results of Barlow and Verrillo. The ordinates of the binocular data are about twice the ordinates of the monocular ones in approximate agreement with perfect arithmetic summation. Thus, the law of additivity is conserved in vision for ganzfeld illumination.

Chemical senses – olfaction and gustation, do not appear to be subject to arithmetic additivity in the sense that hearing, touch and vision in ganzfeld illumination are. According to authoritative studies, the additivity that can be found is incomplete in most instances, but, in some instances, there may be what has been called "hyperadditivity" – the sum exceeds arithmetic addition.

With respect to olfaction, Cain (1977) found by magnitude estimation for n-butyl alcohol that an odorant delivered to both nostrils appears stronger than when delivered to one only. The ratio does not depend on odor intensity, in agreement with arithmetic additivity. However, it is smaller than two. The reduced ratio could mean partial addition, but Cain showed that it is due most probably to adaptation. A preceding stimulus delivered to one nostril makes the following stimulus appear less intense in either nostril. Accordingly, addition is preceded by a mutual interaction in contradiction of the law of additivity. Burglund and Olsson (1993) found a similarly incomplete addition between odorants that are different from each other. Here too, the deficit in additivity was approximately independent of odorant intensity.

Incomplete additivity is also found in gustation, but in many instances hyperadditivity takes place. The latter often occurs in the sweetness of sugars, as found for example by Moskowitz (1973) who used the taste and spit method and normalized magnitude estimation. The effect may have been exaggerated by the method. Bartoshuk and Cleveland (1977) who applied a similar psychophysical method, found that it is smaller, if present at all, when a flow method is used in which the substance to be tested is made to flow over the tongue. With this method, the sweetness of sugars increased with their concentration according to power functions with exponents approximating unity, and the sweetness magnitudes of sugar mixtures appeared to be additive. However, the direct proportionality associated with the unity exponent prevented an unequivocal conclusion that true additivity took place. Otherwise, Bartoshuk and Cleveland did not find the results compatible with arithmetic additivity with either method. In general, gustatory substances are likely to interact with each other before summation of their subjective magnitudes, and the presence of one substance tends to change the taste and apparent intensity of another substance with a different taste. There are exceptions, however. They are encountered in bitter substances. For example, adaptation to QHCL does not reduce the bitterness of urea; neither adaptation to QHCL nor to urea reduces the bitterness of PTC; adaptation to caffeine does not reduce the bitterness of urea and vice versa (Bartoshuk and Cleveland, 1977). Perhaps in these cases, arithmetic additivity of subjective taste magnitudes is possible. Otherwise, it appears to be questionable in chemical senses.

References

Barlow, R.B. and Verrillo, R.T. (1976). Brightness sensation in a ganzfeld. *Vision Research*, 16; 1291–1297.

Bartoshuk, L.M. and Cleveland, C.T. (1977). Mixtures of substances with similar tastes: A test of a physiological model of taste mixture interactions. *Sensory Processes*, 1; 177–186.

Bolanowski, S.J. (1987). Contourless stimuli produce binocular brightness summation. *Vision Research*, 27(11); 1943–1951.

Bolanowski, S.J., Checkowsky, C.M., and Wegenack, T.M. (1993). And now, for our two senses. In: *Sensory Research: Multimodal Perspectives*. R.T. Verrillo (Ed.), Erlbaum, Hillsdale, NJ, pp. 211–231.

Burglund, B. and Olsson, M.J. (1993). Odor-intensity interaction in binary and ternary mixtures. *Perception & Psychophysics*, 53(5); 475–482.

Cain, W.S. (1977). Bilateral interaction in olfaction. *Nature*, 268; 50–52.

Caussé, R. and Chavasse, P. (1942). Différence entre l'écoute binauriculaire et monauriculaire pour la perception des intensités supraliminaire. *Comptes Rendus de la Societe de Biologie*, 136; 405–406.

Fletcher, H. (1929). *Speech and Hearing*. D. Van Nostrand Co., New York.

Fletcher, H. (1935). Newer concepts of the pitch, the loudness and the timbre of musical tones. *Journal of the Franklin Institute*, 220; 405–429.

Fletcher, H. and Munson, W.A. (1933). Loudness, its definition, measurement and calculation. *Journal of the Acoustical Society of America*, 5; 82–108.

Hellman, R.P. and Zwislocki, J.J. (1961). Some factors affecting the estimation of loudness. *Journal of the Acoustical Society of America*, 33; 687.

Hellman, R.P. and Zwislocki, J.J. (1963). Monaural loudness function at 1000 cps ans interaural summation. *Journal of the Acoustical Society of America*, 35(6); 856–865.

Hellman, R.P. and Zwislocki, J.J. (1968). Loudness determination at low sound frequencies. *Journal of the Acoustical Society of America*, 43(1); 60–64.

Levelt, W.J.M., Riemersma, J.B., and Bunt, A.A. (1972). Binaural additivity of loudness. *British Journal of Mathematical and Statistical Psychology*, 25; 51–68.

Luce, R.D. and Tukey, J.W. (1964). Simultaneous conjoint measurement: A new type of fundamental measurement. *Journal of Mathematical Psychology*, 1; 1–27.

Marks, L.E. (1978). Binaural summation of the loudness of pure tones. *Journal of the Acoustical Society of America*, 64(1); 107–113.

Marks, L.E. (1979). Summation of vibrotactile intensity: An analog to auditory critical bands? *Sensory Processes*, 3; 188–203.

Marks, L.E., Algom, D., and Benoit, J.-P. (1991). Dichotic summation of loudness with small frequency separations. *Bulletin of the Psychonomic Society*, 29(1); 62–64.

Moskowitz, H.R. (1973). Models of sweetness additivity. *Journal of Experimental Psychology*, 99(1); 88–98.

Scharf, B. and Fishkin, D. (1970). Binaural summation of loudness reconsidered. *Journal of Experimental Psychology*, 86; 374–379.

Shaw, W.A., Newman, E.B., and Hirsh, I.J. (1947). The difference between monaural and binaural thresholds. *Journal of Experimental Psychology*, 37; 229–242.

Stevens, S.S. and Davis, H. (1938). *Hearing, It's Psychology and Physiology*. John Wiley & Sons, Inc., New York.

Verrillo, R.T. (1968). A duplex mechanism of mechanoreception. In: *The Skin Senses*. D.R. Kenshalo (Ed.), C.C. Thomas, Springfield, IL, pp. 139–159.

Verrillo, R.T. and Gesheider, G.A. (1975). Enhancement and summation in the perception of two successive vibrotactile stimuli. *Perception & Psychophysics*, 18; 128–136.

Zwicker, E., Flottorp, G., and Stevens, S.S. (1957). Critical bandwidth in loudness summation. *Journal of the Acoustical Society of America*, 29; 548–557.

Zwislocki, J.J. (1983). Group and individual relations between sensation magnitudes and their numerical estimates. *Perception & Psychophysics*, 33; 460–468.

Zwislocki, J.J., Ketkar, I., Cannon, M.W., and Nodar, R.H. (1974). Loudness enhancement and summation in pairs or short sound bursts. *Perception & Psychophysics*, 16(1); 91–95.

Zwislocki, J.J. and Sokolich, W.G. (1974). On loudness enhancement of a tone burst by a preceding tone burst. *Perception & Psychophysics*, 16(1); 87–90.

Chapter 4
General Law of Differential Sensitivity

4.1 Introduction

In classical psychophysics, Weber's law refers to the ratios between the just noticeable increments in stimulus intensity and the stimulus intensity to which they are
added. In more general terms, the increments are designated as just noticeable differences and abbreviated as "jnd's." Intuitively, one would expect that the size of a jnd
should depend on the rate at which the subjective magnitude of the percept produced
by a stimulus increases with the intensity of the stimulus. More recent research has
demonstrated that the expectation is fulfilled only in association with specific experimental methods. Perhaps surprisingly, other methods indicate an independence of
the jnd of the rate of growth of the response magnitude. I was able to develop a
mathematical equation that describes the jnd relationships for both instances, those
in which the jnd depends on the rate of growth and those in which it does not. The
equation includes Weber's law but adds to the dimension of stimulus intensity the
dimension of the rate of growth of the response magnitude. I call it the "general law
of differential sensitivity."

As an introduction to the general law, I define Weber's law and give some examples of its applications, beginning with the experiments on lifted weights, which led
E.H. Weber to the observation that the just noticeable difference between two lifted
weights is directly proportional to the magnitude of the weights,

$$\Delta w = cw \tag{4.1}$$

or

$$\frac{\Delta w}{w} = c \tag{4.2}$$

where Δw means the just noticeable weight increment, w, the smaller or standard
weight, and c, a constant. The latter equation is designated as Weber's fraction.

Weber's fraction is often written in a logarithmic form and identified as the jnd; in
general terms: $\log(\Delta I/I)$ and also $\log(1 + \Delta I/I)$, where I means stimulus intensity.

The results of an experiment on weight jnds described by Engen (1972) are illustrated in Fig. 4.1. Two observers participated in the experiment, and the figure shows

J.J. Zwislocki, *Sensory Neuroscience: Four Laws of Psychophysics,*
DOI: 10.1007/978-0-387-84849-5_4,
© Springer Science+Business Media LLC 2009

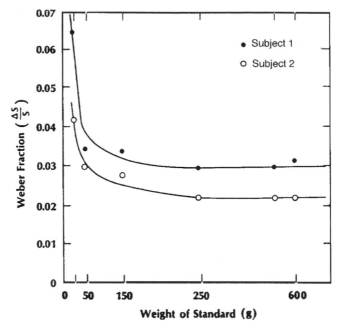

Fig. 4.1 Individual Weber fractions for lifted weights as functions of weight, obtained with the method of constant stimuli. The standard deviation of the responses was accepted as the jnd modified from Engen (1972), with permission granted from copyright holder

the individual results they produced by the method of constant stimuli. Note that one observer produced smaller jnds (smaller Weber fractions) than the other and that the fractions remained approximately constant (Weber's Law) for sufficiently large weights. As the weights decreased below a certain value, Weber' fractions gradually increased, so that Weber's Law was no longer satisfied. The departure from Weber's Law at low stimulus magnitudes appears to hold for all sense modalities. As a consequence, a correction factor has been introduced to Weber's fraction in the form of a small constant. The corrected fraction takes the form:

$$\frac{\Delta w}{(w+a)} = c \tag{4.3}$$

Another classical example of Weber's-law validity can be found in experiments on discrimination of light intensity performed by König and Brodhun (1889). The stimulus consisted of a circular disk of light split in the middle in two halves of equal luminosity, as shown in Fig. 4.2. The observers were fixating at the center of the disc. While they did so, a flash of light of variable intensity was added to one half of the disc. The task of the observers was to detect the flash. The minimum intensity at which the flash could be detected was accepted as the differential threshold. The threshold is plotted in Fig. 4.3 by two solid lines corresponding to scotopic and photopic vision, respectively, according to Hecht (1934). The unfilled circles were

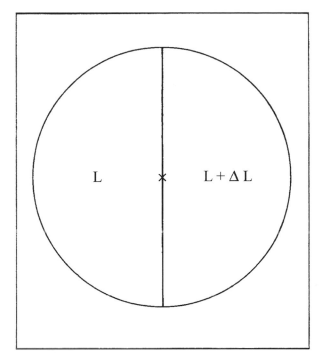

Fig. 4.2 Achromatic luminous disc split in the middle for determination of light jnds. The observers fixated the cross in the middle, and short light flashes were added to the right side of the disc modified from Riggs (1972)

generated by Koenig, the filled ones by Brodhun. Note, that Weber's fraction, $\Delta L/L$, is almost constant at sufficiently high luminosities and increases sharply toward the visual threshold, according to the modified fraction, $\Delta L/(L+a)$.

Weber's Law holds also in audition for stimuli with broad-band spectra, like broad-band random noise or square waves of irregular length, as shown in Fig. 4.4 according to the experiments of Miller (1947). Again, Weber's fraction is constant at sufficiently high stimulus intensities but increases near the absolute threshold of audibility. For pure tones, however, Weber's fraction never reaches a horizontal asymptote but continues to decrease slowly. The phenomenon has been called "near miss to Weber's Law" by McGill and Goldberg (1968). The difference between jnd characteristics obtained with broad-band noise and complying with Weber's Law and jnd characteristics obtained on the same group of five observers with a 4-kHz pure tone and complying with the near miss is shown in Fig. 4.5. More extensive data, demonstrating that the near miss occurs generally at all audible pure-tone frequencies are reproduced in Fig. 4.6 (Ozimek and Zwislocki, 1996). The near miss seems to be present in vibrotaction for both sinusoidal and random-noise stimuli, as is evident in Fig. 4.7 for two stimulus paradigms. The lack of a difference suggests that they are analyzed in a different way than in hearing.

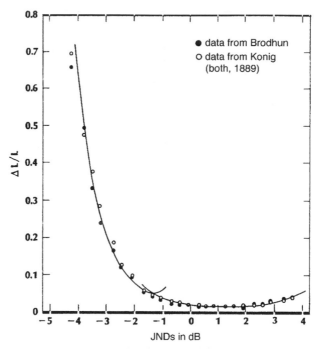

Fig. 4.3 Differential threshold for achromatic light as a function of light intensity obtained with stimulus configuration shown in Fig. 4.2. Unfilled circles show individual data obtained by König, filled circles, those by Brodhun (1889). Solid lines show interpolating curves for scotopic and photopic vision according to Hecht. Reproduced from Riggs (1972), after Hecht (1934)

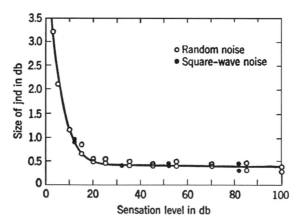

Fig. 4.4 Differential thresholds for white noise presented binaurally (unfilled circles) and for irregular rectangular waves (filled circles) as functions of sound intensity. Reproduced from Licklider (1951), after Miller (1947), with permission from John Wiley & Sons, Inc.

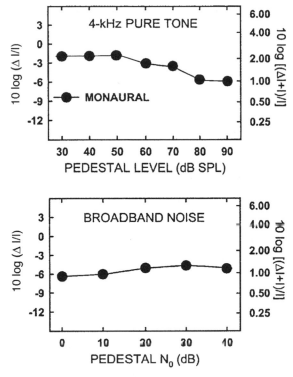

Fig. 4.5 Mean differential thresholds for a 4-k Hz tone and broad-band noise as functions of sound intensity, obtained on the same group of five listeners. Modified from Stellmack, Stellmack, Viemeister and Byrne (2004), reproduced with permission from the American Institute of Physics

Mathematically, the near miss can be expressed by replacing the constant c in Eq. 4.1 with the expression $Cw^{-\mu}$, or, in more general terms, $CI^{-\mu}$. Then, the equation becomes:

$$\frac{\Delta I}{I} = CI^{-\mu} \tag{4.4}$$

A similar expression was used by McGill and Goldberg (1968), Schacknow and Raab (1973), Penner et al. (1974), Jesteadt et al. (1977) and some others.

4.2 Weber's Fraction Independent of the Rate of Response Growth

Clinical studies of Weber's fraction on patients with monaural hearing loss associated with the phenomenon of loudness recruitment revealed surprisingly that the numerical value of the fraction remained approximately constant when the loudness was kept constant, independent of the rate of loudness growth (Lüscher and

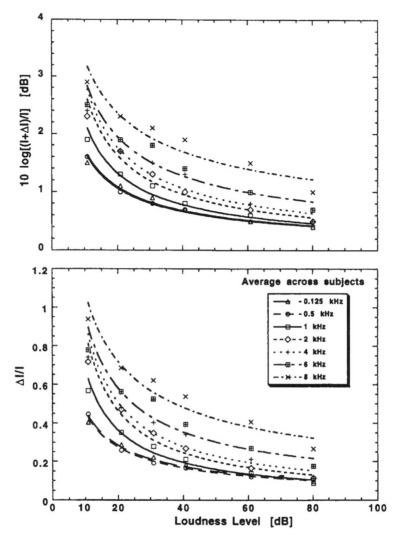

Fig. 4.6 Mean differential thresholds for pure tones at several frequencies as functions of sound intensity, plotted on logarithmic and linear ordinate scales. Reproduced from Ozimek and Zwislocki (1996), with permission from the American Institute of Physics

Zwislocki, 1948, 1951). The loudness constancy was obtained by interaural loudness matching which resulted in inequality of associated sound intensities. Thus, the conclusion was reached that the intensity jnd was associated with loudness rather than stimulus intensity. The coupling with loudness led to indirect determination of loudness recruitment, symptomatic of hearing loss that originates in inner-ear pathology – the "Lüscher – Zwislocki method."

The results of Lüscher and Zwislocki were subsequently confirmed by laboratory experiments of Jordan (1962; Zwislocki and Jordan, 1986) on the occasion of his

Fig. 4.7 Mean differential thresholds for vibrotaction obtained at two vibration frequencies and narrow-band noise with two methods of stimulus presentation – gated pedestal and continuous pedestal. The data are plotted on a logarithmic scale as functions of sensation level. Reproduced from Gescheideer et al. (1990), with permission from the American Institute of Physics

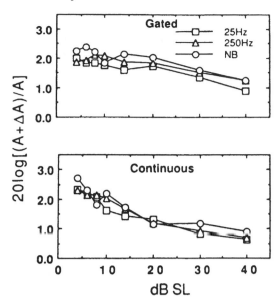

doctoral-dissertation. As in the studies of Lüscher and Zwislocki, the stimuli consisted of trapezoidally amplitude-modulated pure tones with a modulation rate of $2.5\,s^{-1}$. According to the classical studies of Riesz (1928), this modulation rate was expected to produce the lowest jnd values. The modulation was interrupted at random time intervals to facilitate its detection and check on the listeners' false-positive responses. The listeners responded according to the psychophysical method of limits. The population of listeners with normal hearing consisted of 26 college students, the population with hearing loss, of 14 adults aged 20–73 years with a median of 42 years. Their hearing loss ranged from 30 to 75 dB in the worse ear and did not exceed 10 dB at the test frequencies in the better ear.

Jordan's results obtained at 1 kHz on the group with normal hearing are compared to the results of Riesz (1928) and of Doerfler (1948) and to the more recent results of Jesteadt et al. (1977) in Fig. 4.8. Whereas the former two were obtained with nearly sinusoidal amplitude modulation, the latter resulted from intensity discrimination in burst pairs and an adaptive forced-choice procedure considered to be criterion free. Interestingly, Jordan's results are in the best agreement with the latter, so that his method should be considered more valid than those of the preceding studies. Note that the results conform with the near miss to Weber's law.

Jordan applied the same modulation method to the group with unilateral hearing loss. The mean jnd values he obtained on the ears with hearing loss were compared to those obtained on the contralateral ears without hearing loss at equal loudness. The results obtained at three sound frequencies – 0.25, 1, and 4 kHz, are shown in Figs. 4.9–4.11, respectively. The mean jnd values are plotted for both ears in dB according to the definition $10\,\log(1 + \Delta I/I)$ as functions of SL in the better ear. Except at 10 dB SL, at all three frequencies, the jnd values prevailing in the presence

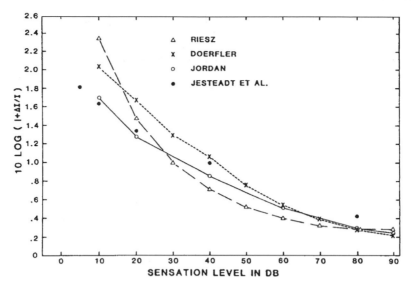

Fig. 4.8 Mean differential threshold for trapezoidal amplitude modulation of a 1-kHz tone as a function of sensation level, compared to preceding studies performed with different methods. Reproduced from Zwislocki and Jordan (1986), with permission from the American Institute of Physics

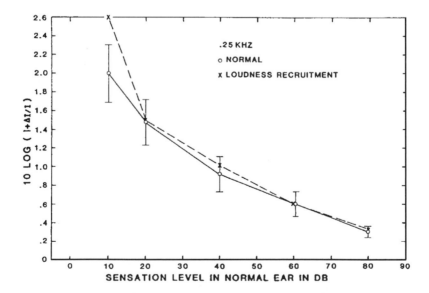

Fig. 4.9 Mean differential thresholds for trapezoidal amplitude modulation of a 0.25-kHz tone as functions of sensation level, obtained on listener groups with normal hearing (solid line) and hearing loss associated with loudness recruitment (intermittent line). Vertical lines indicate double standard deviations of individual responses. Reproduced from Zwislocki and Jordan (1986), with permission from the American Institute of Physics

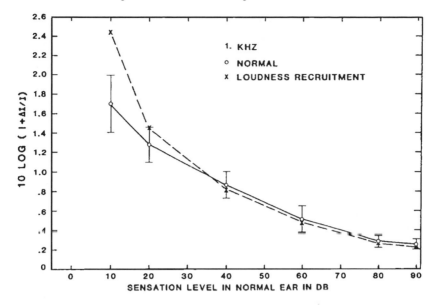

Fig. 4.10 Similar to Fig. 4.9 for a carrier of 1 kHz. Reproduced from Zwislocki and Jordan (1986), with permission from the American Institute of Physics

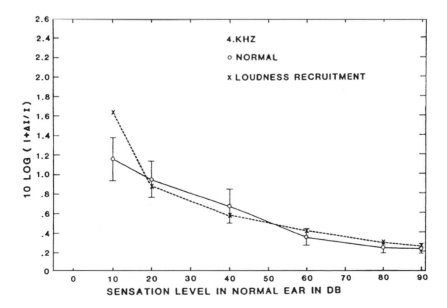

Fig. 4.11 Similar to Fig. 4.10 for a carrier of 4 kHz. Reproduced from Zwislocki and Jordan (1986), with permission from the American Institute of Physics

of hearing loss agree approximately with those found in its absence. Thus, equal loudness magnitudes were associated with equal jnds, except at the lowest SLs, independent of the SL relationships. In other words, the jnds appeared to depend on loudness, a response variable, rather than a corresponding stimulus variable. It is also true that, because of loudness recruitment in the worse ear, the jnds in this ear were associated with a more rapid loudness growth than equal jnds in the better ear. Accordingly, the jnds did not depend on the rate of loudness growth.

The results of Jordan and Zwislocki were confirmed by Stillman et al. (1993) who used more contemporary detection methods. Their experiments were performed on two groups of observers, one consisting of six listeners with normal hearing, the other, of eight listeners with predominantly unilateral hearing loss. In the latter group, five listeners were tested at one sound frequency and three, at two frequencies. In all, 11 sets of data were obtained, 3 at 0.5 kHz, 2 at 4 kHz and 1 each at 1.0, 2.0, 2.5, 3.0 6.0 and 6.5 kHz. The choice of the frequencies was dictated by the distributions of hearing loss. The loudness equality between the better and worse ears was determined by the method of adjustment in which, alternately, the stimuli in the better and worse ears served as reference standards. Every listener made 12 loudness matches at each frequency, six with the standard stimulus in the better ear, and six with the standard stimulus in the worse ear. The stimuli consisted of 200-ms bursts of pure tones produced for both ears by the same generator. They were alternated between the two ears with 500-ms inter-burst time intervals. The results are displayed graphically in Fig. 4.12 in terms of SPLs producing loudness equality between the two ears. The SPL in the worse ear is indicated on the abscissa axis, the SPL in the better ear, on the ordinate axis. The loudness-equality plots are organized in three frequency ranges: 0.5, 1.0–3.0, and 4.0–6.5 kHz. The various symbols indicate the measured loudness-equality pairs of SPLs. The unfilled circles belong to absolute threshold values, and the crosses mark the SPLs at which the jnds were measured. The slanted straight lines indicate the SPL relationships corresponding to bilaterally normal hearing.

The jnds were determined by an adaptive, two-interval, forced-choice procedure (Levitt, 1971). The listeners had to decide which of the two time intervals contained the larger increment in the pedestal intensity, which was randomized between the two intervals. After every two correct responses, the sound intensity of the increment was decreased by 2 dB, after every incorrect response, it was increased by 2 dB. In this way, the jnd was tracked at the level of 70.7% correct responses. The increments were each 200 ms long and were separated by 400-ms intervals. The pedestals started 400 ms before the first increment and ended 400 ms after the second one, so that they lasted for 1,600 ms. According to Turner et al. (1989) this pattern produces the lowest jnds. To avoid audible transients, the pedestals were turned on and off with linear rise and fall times of 25-ms duration; the increments were gated with cosine-square skirts having 20-ms time constants. The sequences were initiated by the listener's responses. Each block of responses was terminated after 14 reversals of intensity changes, and each jnd estimate was the average of 12 blocks.

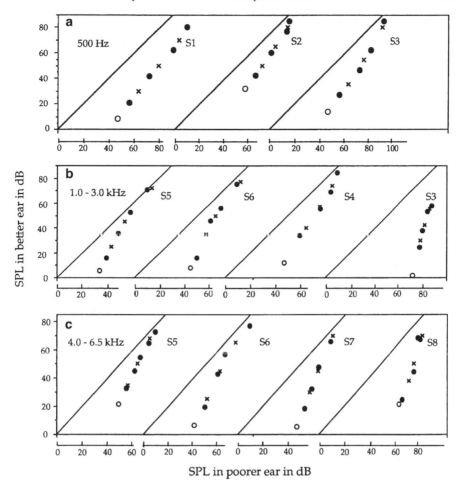

Fig. 4.12 Individual loudness-level functions of listeners with monaural hearing loss at the test frequency in three frequency bands. Filled circles indicate loudness-matching data; unfilled circles – the thresholds of audibility; crosses – the levels at which jnds were measured. Reproduced from Stillman et al. (1993), with permission from the American Institute of Physics

Individual results for both ears at loudness equality, plotted in terms of $10 \log(1 + \Delta I/I)$, are displayed in Fig. 4.13 as functions of SL in the better ear for the same frequency ranges as the loudness-level functions of Fig. 4.12. The filled circles belong to the better ear, the unfilled ones to the worse ear. The vertical bars indicate the standard deviations of the jnd estimates. Although the variability of the estimates is substantial, especially at higher sound frequencies, on the whole, the jnds tend to be equal in both ears, confirming on the individual basis the earlier findings made on group averages that the intensity jnds tend to be equal at equal loudness, independent of the slope of the loudness function. In greater detail, of the 33 pairs of data points in Figs. 4.13 and 4.15 indicate no significant interaural

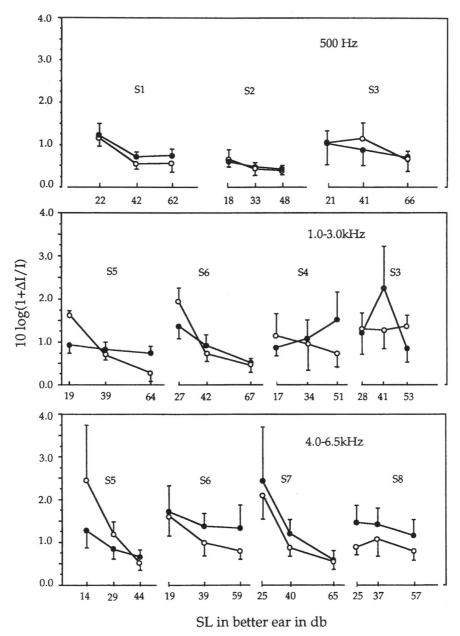

Fig. 4.13 Individual mean intensity jnds corresponding to loudness level functions of Fig. 4.12 (crosses) and associated standard deviations indicated by vertical line segments. Filled circles belong to the better ear, unfilled circles, to the poorer ear at loudness equality. The data are plotted as function of SL in the better ear. Reproduced from Stillman et al. (1993), with permission from the American Institute of Physics

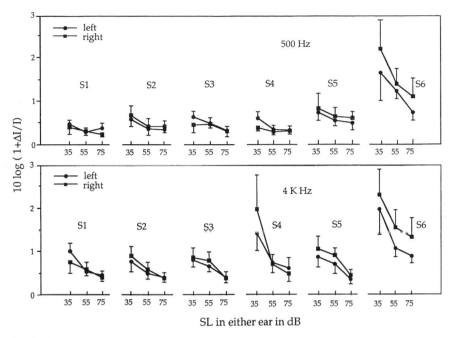

Fig. 4.14 Individual mean intensity jnds and corresponding standard deviations determined at equal loudness in each ear of listeners with binaurally normal hearing. The jnds were measured at two sound frequencies and are plotted as functions of SL in either ear. Reproduced from Stillman et al. (1993), with permission from the American Institute of Physics

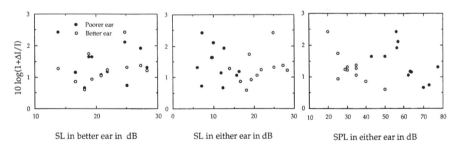

Fig. 4.15 Scatter plots of jnd values obtained at equal loudness in the better (*unfilled circles*) and poorer (*filled circles*) ears, shown relative to the SL in the better ear corresponding to loudness equality with the poorer ear (*left panel*), relative to the corresponding SL in either ear (*middle panel*), and relative to the corresponding SPL in either ear. Reproduced from Stillman et al. (1993), with permission from the American Institute of Physics

difference, 5, a significantly smaller jnd in the better ear, and 13, in the worse ear. To check if the monaural hearing loss increased the interaural asymmetry in jnd values, the jnds were measured at two sound frequencies – 0.5 and 4 kHz, in both ears of six listeners with binaurally normal hearing. Somewhat surprisingly, significant individual asymmetries came out to be even larger than in the population with monaural hearing loss, as is evident in Fig. 4.15. However, the increased differences can be ascribed to a smaller monaural variability (standard deviation) of the data.

To further test the independence of intensity jnds of the slope of the loudness-level functions, Stillman et al. plotted the individual interaural ratios of jnd values obtained at equal loudness magnitudes in both ears of the listeners with monaural hearing loss in relation to the slopes of the corresponding interaural loudness-level functions. If the jnds were independent of the slope, their average ratios should average to unity. Computation of the empirical data produced ratios of 1.08, 0.89, and 0.98 for the low, moderate and high SLs, respectively, producing the grand average of 0.983, insignificantly different from unity.

Because loudness increases with SPL and SL, Stillman et al. found it necessary to demonstrate that the jnds they obtained were not correlated with these variables but only with loudness. To this end, they made three plots, one, relating the jnds obtained in both ears at equal loudness values to the corresponding SLs in the better ear, one, relating the jnds obtained in each ear to the SLs at which they were obtained in the respective ear, and one, relating the jnds to the respective SPLs. The plots are shown in Fig. 4.15. Clearly, the jnds obtained at equal loudness magnitudes extend over the same range of SLs in the better ear, but the jnds not obtained at equal loudness magnitudes but in relation to SLs or SPLs of each ear cover separate ranges of these variables. Thus, as paradoxical as this may appear, the intensity jnds are not tied to the stimulus variables SL or SPL but to a response variable – loudness.

To relate the results of Stillman et al. more directly to Weber fractions, the intensity jnds they obtained in both ears at equal loudness magnitudes are plotted in the form of $\Delta I/I$ versus the SLs in the better ear in Fig. 4.16. The practically complete overlap of the two jnd populations is evident. Interestingly, the jnds of the worse ear decrease together with those of the better ear, as the SL along with the loudness increase, in agreement with the near miss to Weber's law. Accordingly, hearing loss does not appear to affect the near miss, when the jnds are plotted as a function of loudness level.

In an elegant experiment, Hellman et al. (1987) managed to determine intensity jnds while keeping both the SL and loudness level (LL) constant and varying only the slope of the LL curve. They achieved this by embedding 1 kHz tone bursts in a narrow-band and a wide-band noise, respectively. Masking by the narrow-band noise produced a substantially steeper LL curve than masking by the broad-band noise. Choosing the noise intensities appropriately, they were able to make the LL curves cross at one point. The measurements were performed on seven listeners with the loudness balance method for the LL curves and on five listeners with an adaptive two-interval forced-choice method for the jnds. The chosen response schedule converged on 75% correct responses. Their mean results are shown in Fig. 4.17 by means of the curves for the LL and a block diagram for the jnds measured at the crossing point. Clearly, the jnds are about the same in the presence of the narrow-band noise producing the steeper LL curve, and the wide-band noise, producing the less steep LL curve. Thus, the independence of the intensity jnds of the slope of the LL and, therefore, loudness curves was confirmed.

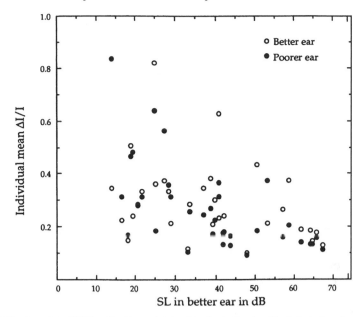

Fig. 4.16 Scatter plot of Weber fractions obtained in the better (*unfilled circles*) and poorer (*filled circles*) ears at loudness equality relative to the SL in the better ear corresponding to loudness equality with the poorer ear. Reproduced from Stillman et al. (1993), with permission from the American Institute of Physics

Fig. 4.17 Crossing loudness-level curves obtained at 1 kHz in the presence of a narrow-band, respectively, broad-band noise. Circles on the abscissa axis indicate the respective thresholds of audibility. The block heights in the inset indicate the Weber-fraction values obtained at the crossing point of the curves, in the presence of the narrow-band, respectively, broad-band noise. The thick line segments show the corresponding standard deviations of the listener responses. Reproduced from Hellman et al. (1987), with permission from the American Institute of Physics

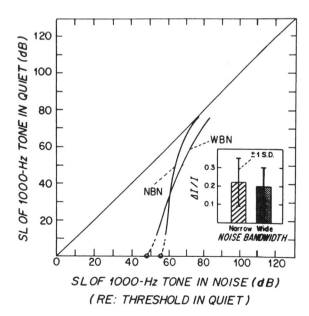

A further confirmation of the jnd independence of the slope of the LL functions was obtained by Johnson et al. (1993) who performed experiments similar to those of Hellman et al. (1987). However, they also measured the jnds at equal LL outside the crossing points and focused their attention on individual results obtained on three listeners. Also, using two SPL levels for the narrow-band noise, they were able to obtain two crossing points between the LL curves in the wide-band and narrow-band masking noises. Further, the LL curves were obtained by the adaptive procedure of Jesteadt (1980) maximizing the accuracy of the obtained loudness balances. The jnd experiments were performed by the adaptive two-interval forced-choice procedure converging on 70.7% correct responses, as described by Levitt (1971). The two intensity increments were 1,200 ms long to match the duration of the tone bursts used in the LL experiments and were added to a quasi-continuous pedestal that begun 400 ms before the first increment and ended 400 ms after the second increment. They were separated by 400-ms time intervals. The stimuli were turned on and off with 25-ms ramps, with the exception of the increments for the jnd measurements, which were turned on and off with 10-ms transients. As an example, the LL curves obtained on one listener are shown in Figs. 4.18a and b for the two crossing points. The curve joining the unfilled squares was determined without masking. The jnd results for the crossing points, at which both the loudness magnitudes and the SPLs were equal, are summarized in Fig. 4.19 for all three listeners. The jnds obtained in narrow-band noise (steeper LL curve) are plotted relative to the jnds obtained in the wide-band noise (less steep LL curve). The diagonal intermittent line shows perfect jnd equality in both noise bandwidths. With the exception of one idiosyncratic point, the empirical jnds fall close to this line, indicating independence of the slope of the LL curves. A small tendency of the jnds to be larger in the narrow-band noise may be ascribed to greater amplitude fluctuations in this noise. Equal-loudness jnds, some of which coincide with the crossing points and some that do not, are plotted for one listener and the three conditions – masking by wide-band noise, masking by narrow-band noise, and no noise, as functions of LL in Fig. 4.20. There is no significant difference between the jnds in the three conditions, neither is there a difference between the jnds determined at the crossing points and those that were not. Similar results were obtained for the other listener who completed the experiments.

The results obtained by Johnson et al. confirm the conclusions reached on the basis of the results of the other groups of experimenters, which were discussed above. Paradoxically, intensity jnds depend on loudness, a parameter of the response domain, not on the rate of loudness growth with stimulus intensity ore the stimulus intensity itself. How is this possible?

First, we should realize that an intensity increment is perceived as a loudness increment (Zwislocki and Jordan, 1986; Zwislocki, 1996a, b). Let us call it ΔL. Defining its detectability in terms of the theory of signal detectability transferred from the stimulus domain to the response domain, we can write:

$$d_L' = \frac{\Delta L}{\sigma_L} \tag{4.5}$$

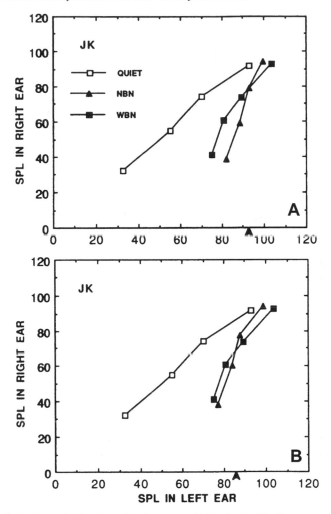

Fig. 4.18 Individual crossing loudness-level curves at 1 kHz obtained in the presence of a narrow-band noise (*filled triangles*) at two SPLs (80 dB in the upper panel and 75 dB in the lower panel), respectively, and a wide-band noise at 77 dB SPLs (*filled squares*). For comparison, the unfilled squares and the interpolating line indicate the loudness-level curves obtained in the absence of masking noise. The arrowheads at the abscissa axis show the SPL values of the intersection points. Reproduced from Johnson et al. (1993), with permission from the American Institute of Physics

where d_L' means the detectability index in the response domain and σ_L, the standard deviation of the neural noise at the level of loudness perception. To find a connection between loudness and stimulus intensity, we can express loudness, L, as a differentiable function of sound intensity,

$$L = F(I) \tag{4.6}$$

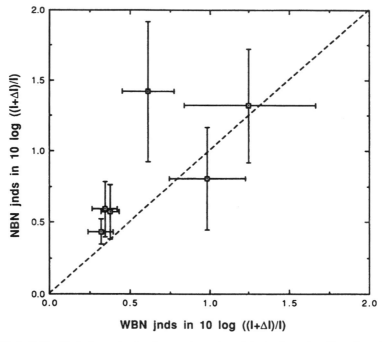

Fig. 4.19 Individual jnds determined at the crossing points of the loudness-level functions obtained in the presence of the narrow-band, respectively, wide-band noises. The jnds determined in the presence of the narrow-band noise are plotted versus the jnds determined in the presence of the wide-band noise. The crossing vertical and horizontal line segment indicate corresponding standard deviations of the listener responses. The intermittent diagonal line indicates perfect jnd independence of the slope of the loudness-level curves (narrow-band versus wide-band noise). Reproduced from Johnson et al. (1993), with permission from the American Institute of Physics

Fig. 4.20 Equal-loudness jnds determined as functions of loudness level in quiet, in the presence of wide-band masking noise, respectively, narrow-band noise. Individual data. Reproduced from Johnson et al. (1993), with permission from the American Institute of Physics

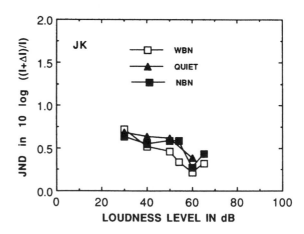

Differentiation of the latter equation, leads to its difference form,

$$\Delta L = F'(I)\Delta I \tag{4.7}$$

Substituting for ΔL its expression of Eq. 4.5 we obtain:

$$d_L'\sigma_L = F'(I)\Delta I \tag{4.8}$$

According to the theory of signal detectability, d_L' is in the vicinity of unity near the detection threshold, so that σ_L should be of the same order of magnitude as ΔL. If it originates in the neural noise associated with stimulus intensity, I, and having a standard deviation, σ_I, of the same order of magnitude as ΔI, it should follow approximately a similar function of stimulus intensity,

$$\sigma_I = a\, F'(I)\sigma_I \tag{4.9}$$

Where a is a proportionality constant independent of I. Division of both sides of Eq. 4.8 by both sides of Eq. 4.9 produces:

$$a\, d_L' = \frac{\Delta I}{\sigma_I} \tag{4.10}$$

an equation free of $F'(I)$, in other words, independent of the slope of the loudness function. Rearranging its terms and dividing both sides by I, we obtain:

$$a\, d_L'\frac{\sigma_I}{I} = \frac{\Delta I}{I} \tag{4.11}$$

An equation for the Weber fraction, that is independent of the slope of the loudness function and of stimulus intensity, I, if σ_I/I is constant, $a\, d_L'$ being independent of I by definition. If σ_I is less than directly proportional to I, an expression for the near miss to Weber's law results. It is possible to write for example $\sigma_I \sim I^{1-\mu}$, as was done by Zwislocki and Jordan (1986). Then, with a numerical modification of constant a

$$\frac{\Delta I}{I} = a\, d_L' I^{-\mu} \tag{4.12}$$

Weber's fraction usually becomes large near the absolute threshold of detectability. The increment is very likely due to a spontaneous internal noise at the sensory periphery. This can be expressed mathematically by adding a constant, σ_{Io}, to the variable σ_I in Eq. 4.11. We then obtain:

$$\frac{\Delta I}{I} = \frac{a\, d_L'(\sigma_I^2 + \sigma_{Io}^2)^{1/2}}{I} \tag{4.13}$$

The latter equation is consistent with the Weber-fraction curves of Figs. 4.1 and 4.3, 4.4. An analogous mathematical expression for the near-miss curves can be obtained by including the constant σ_{Io} in the derivation of Eq. 4.12. Examples of such curves for hearing are given in Figs. 4.5 (upper panel), 4.6, 4.8–4.11.

A mathematical explanation of the apparent paradox according to which Weber's fractions do not depend on the rate of growth of subjective magnitudes as functions of stimulus magnitudes has been developed above for loudness. The assumptions on which it has been based and the structure of the equations involved have a more general character, however.

Outside hearing, the paradox has already been demonstrated in the sense of touch. Gescheider et al. (1994) applied a vibrotactile sinusoidal stimulus to the thenar eminence of the right hand and changed the rate of growth of its subjective magnitude with the vibration amplitude by masking consisting of random-vibration. The test stimulus as well as the masker were delivered through a cylindrical contactor of 2.9-cm^2 contact area, pressed into the skin surface to a depth of 0.5 mm. The vibration of the skin surface was limited to the immediate vicinity of the contactor by a rigid surround. The contactor protruded through a circular opening in the surround, leaving an annular gap of 1 mm width. The test stimulus was kept at a vibration frequency of 250 Hz and was presented in bursts of 700-ms duration. The masker consisted of random noise with a spectrum extending from 37 to 600 Hz. It was presented in 1,500-ms bursts centered on the test-stimulus ones. Because the noise bursts were longer than the sinusoidal test bursts, the observers were able to easily distinguish the latter from the former and to judge their sensation magnitudes. The magnitudes perceived in the presence of noise were compared to the magnitudes perceived in its absence to obtain curves of sensation equality, or sensation-level curves. Several curves were obtained by changing the masking intensity in steps. The experimental results are shown in Fig. 4.21. Three phenomena related to the masker intensity should be noted. As the masker intensity increases, the threshold of detectability increases and, with it, the slope of the sensation-level curves, whereas the sensation level decreases. By cutting the set of the curves of Fig. 4.21 horizontally, it is possible to obtain equal levels of sensation magnitudes associated with different SLs of the 250-Hz stimulus.

Prior to the sensation-level measurements absolute thresholds of detectability were measured for a similar stimulus configuration by means of a two-interval forced-choice adaptive procedure converging on 75% correct responses (Zwislocki et al., 1958; Zwislocki and Relkin, 2001). The same method was used for the measurement of intensity jnds. The test bursts were presented in pairs at several reference (pedestal) levels, one of the test bursts in each pair selected at random having a greater vibration amplitude than the other. The observers had to decide which test burst in a pair was the stronger. The resulting jnds were expressed in dB according to the formula $20\log[(\Delta A + A)/A]$, where A means vibration amplitude and ΔA, the amplitude increment. They were determined as functions of the sensation level of the 250-Hz stimulus, as functions of its amplitude, and as functions of its subjective magnitude level, in the presence and absence of masking. The jnds were found to be invariant under all masking conditions only when the level of the sensation magnitude was constant, as is evident in Fig. 4.22, where the jnds are plotted as functions of the matching SLs of the pedestals presented in the absence of masking. The invariance means that the jnds were dependent on the level of the subjective magnitude rather than on SL (near miss to Weber's Law) and were independent of the rate of sensation growth.

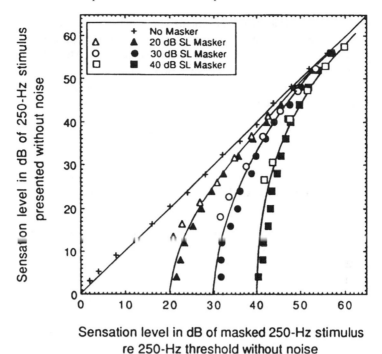

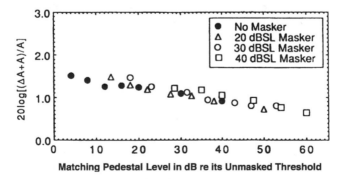

Fig. 4.21 Equal-sensation curves obtained by matching the sensation magnitude of 250-Hz vibration in the presence of masking-noise vibration to 250-Hz vibration in the absence of noise. Unfilled symbols correspond to the variation of the 250-Hz vibration amplitude in the absence of masking noise, filled symbols correspond to the variation of the 250-Hz vibration in the presence of masking noise. Crosses correspond to sensation-magnitude matching in the absence of noise masking. Reproduced from Gescheider et al. (1994), with permission from the American Institute of Physics

Fig. 4.22 Equal-sensation magnitude jnds for 250-Hz vibration, obtained in the presence of several levels of masking noise and without masking, plotted as functions of SL referred to the threshold of detectability in the absence of masking noise. Masking increased the slope of the sensation-magnitude curves. Reproduced from Gescheider et al. (1994), with permission from the American Institute of Physics

The results of the vibrotactile experiments support the conclusion that Weber's fractions are independent of the rate of growth of subjective magnitudes as functions of stimulus magnitudes in many if not all sensory modalities. They depend on the subjective magnitudes themselves rather than on the magnitudes of physical stimuli that underlie them. Nevertheless, the following section shows that the independence of the jnds of the rate of growth of subjective magnitudes only holds for intensity increments, not for the jnds derived from statistical errors in matching subjective magnitudes.

4.3 The Generalized Law of Differential Sensitivity

When the subjective magnitudes of two psychological variables are matched to each other so as to appear equal, one is held constant as a reference standard, the other is varied until a match is obtained. The operation is performed by manipulating the underlying stimuli. The match is never perfect but is distributed statistically around the point of equality. The distribution is reflected in the distribution of the underlying stimulus values. The standard deviation of the latter is taken as a measure of the variability of the subjective magnitudes. In a counterbalanced procedure, the magnitudes of both variables are alternately used as standards. When their underlying stimuli are given successively different values, stimulus curves of subjective magnitude equality are obtained. By varying the functional stimulus-response relationships, families of such curves with varying slopes can be generated.

When experimenters used the standard deviations as measures of jnds, they noticed that the jnds depended on the slopes of the matching curves, in apparent contradiction of the jnd independence of the growth rate of the subjective magnitudes described in the preceding section (Cefaratti and Zwislocki, 1994, for review). An auditory example is illustrated in Fig. 4.23. The stimuli, consisting of 1 kHz tone bursts, 500 ms in duration, were repeated at a rate of $1\,\mathrm{s}^{-1}$ and presented alternately to both ears. To avoid audible on and off transients, they were turned on and off with 50-ms temporal ramps. The tone intensity was kept constant in one ear at several consecutive SLs and was adjusted by the listener in the opposite ear to obtain binaural loudness equality. A random masking noise with a spectrum extending from 857 to 1425 Hz was delivered to one ear for the purpose of increasing the threshold of audibility in that ear. The noise intensity was set consecutively at two levels so chosen as to produce 40, respectively, 60-dB threshold shifts. The threshold shifts were associated with steepened loudness-level curves determined by interaural loudness matches and shown by the straight solid lines in the figure. The loudness matches were performed by manipulating round, unmarked knobs. The three listeners participating in the experiments repeated the matches six times at every SL of the reference tone. Standard deviations of the variability of the matches were determined and, because they did not vary systematically with the SL, were averaged over all the SLs involved. The averaged double standard deviations are shown for one listener in Fig. 4.23 by crossing straight-line sections, the vertical line, for intensity

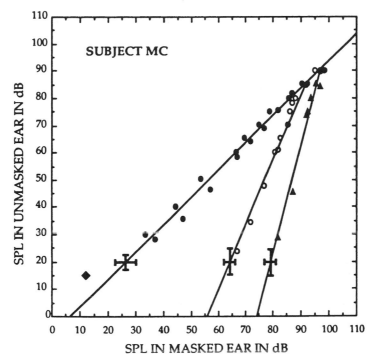

Fig. 4.23 Monaural loudness-level curves obtained at 1 kHz in the absence and presence of two levels of monaural masking noise. The diamond indicates the absolute detection thresholds of both ears in the absence of masking. The data points were obtained by interaural loudness matches obtained by varying the sound intensity consecutively in the masked and unmasked ears. The crossing horizontal and vertical line segments indicate the standard deviations of the matching intensity settings in the masked and unmasked ear, respectively. Reproduced from Cefaratti and Zwislocki (1994), with permission from the American Institute of Physics

adjustments in the unmasked ear, the horizontal line, for those in the masked ear. The filled circles indicate the mean loudness matches obtained in the absence of masking noise, the empty circles and filled triangles, the matches obtained in the presence of two noise levels, respectively. The diamond shows the thresholds of audibility of the two ears in the absence of noise. Note in particular, the lengths of the standard-deviation lines serving as jnd measures. The horizontal line belonging to loudness adjustments in the masked ear becomes somewhat shorter as the slope of the loudness level curve increases. This seems to be consistent with the belief held by some researchers since Fechner that intensity jnds should be inversely proportional to the slope of the magnitude functions. However, the change in line length is much smaller than would be expected on the basis of inverse proportionality. Surprisingly, the standard deviation of the loudness adjustments in the unmasked ear having a constant slope of the loudness function does not remain invariant but increases with the slope of the loudness level function of the contralateral ear. These relationships become more clearly relevant to jnds when Weber fractions are calculated from the standard deviations expressed in dB. This can be done with the help of the familiar equation converting decibels to linear ratios,

$$\frac{\Delta I}{I} = 10^{\sigma/10} \tag{4.14}$$

Where σ means here the standard deviation. Weber fractions calculated from the standard deviations of the loudness adjustments averaged over all three listeners participating in the experiment are plotted as functions of the slopes of the obtained loudness-level functions in Fig. 4.24. Clearly, the standard deviation belonging to loudness adjustments in the masked ear decreases inversely with the slope, when the slope is small but not when it is large. The standard deviation belonging to the unmasked ear increases monotonically with the slope.

To verify that the standard-deviation patterns of Figs. 4.23 and 4.24 are not idiosyncratic and specific to hearing, Cefaratti and Zwislocki (1994) performed an intersensory experiment in which loudness was matched to line length. The loudness variable of the unmasked ear was replaced by a luminous line segment projected on

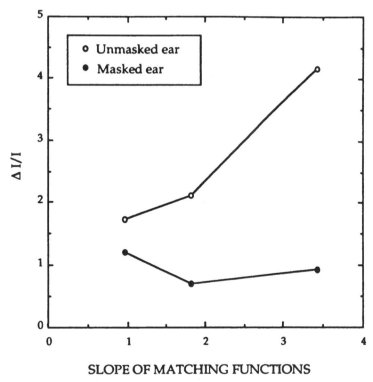

SLOPE OF MATCHING FUNCTIONS

Fig. 4.24 Weber fractions derived from mean standard deviations of loudness-level settings by three listeners, produced in connection with interaural loudness matching, plotted versus the slope of the loudness-level curves. Unfilled circles refer to sound-intensity variation in the unmasked ear, filled circles, to the variation in the masked ear. Reproduced from Cefaratti and Zwislocki (1994), with permission from the American Institute of Physics

the translucent window pan in the sound-proofed booth in which the auditory experiments were performed. The setup permitted the line length to be varied by manipulating the same knob that was used for sound-intensity adjustments. The same three observers participated as in the auditory experiment and the same psychophysical method of adjustment was used. As an example, Fig. 4.25 shows individual data for the line lengths displayed as functions of the SPL of the 1-kHz tone at three masking levels. As in the auditory experiment, the horizontal lines refer to the standard deviations of the tone-intensity adjustments in the masked ear; the vertical lines, to the adjustments of the line length. The relationships confirm those found in the auditory experiment. Increasing rate of loudness growth produced by masking is associated with a decreasing standard deviation of tone-intensity adjustments but with an increasing standard deviation of the matching line-length adjustments. Corresponding Weber fractions averaged over the three observers are shown in Fig. 4.26. Like the auditory standard deviations, the auditory Weber fractions decrease, as the rate of loudness growth increases, by contrast, the matching line-length ones increase.

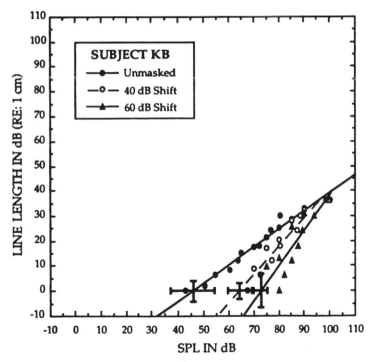

Fig. 4.25 Individual cross-modality matching functions obtained by matching line length to loudness of 1-kHz tone bursts presented to one ear in the absence (*filled circles*) and presence of two levels of narrow-band masking noise (*unfilled circles and filled triangles*). The crossing horizontal and vertical line segments indicate standard deviations of the tone-intensity and line-length settings, respectively, produced in connection with the subjective-magnitude matching. Reproduced from Cefaratti and Zwislocki (1994), with permission from the American Institute of Physics

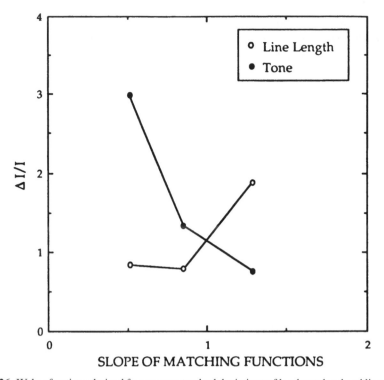

Fig. 4.26 Weber fractions derived from mean standard deviations of loudness-level and line-length settings by three observers, produced in connection with the intermodal magnitude matching, plotted versus the slope of the intermodal matching curves. Unfilled circles refer to line-length variation, filled circles, to tone intensity variation. Reproduced from Cefaratti and Zwislocki (1994), with permission from the American Institute of Physics

The two examples of Weber fractions associated with sensation-magnitude matching, one intrasensory, the other, intersensory, demonstrate that the Weber fractions, unlike those obtained in connection with the detection of small stimulus increments, do depend on the slopes of subjective magnitude functions. They do so in a complicated way not simply related to the growth rate of subjective magnitudes. What is the explanation for these apparently paradoxical relationships?

Perhaps surprisingly, the slope dependence can be accounted for in a manner similar to the way the slope independence has been accounted for. It is rooted in the detection index d' applied to the response level of signal processing rather than to the peripheral level. Only here, two sources of noise associated with two signals have to be introduced. For clarity, let as assume that the two signals consist of two tones, each presented to a different ear, and the spectral noise bands are sufficiently narrow to be processed according to approximately the same functions as the signals. Because of the well-known phenomenon of binaural summation described in the preceding chapter, we assume that the variances of the two noises sum, so that the total noise standard deviation amounts to

$$\sigma_L = (\sigma_{L1}^2 + \sigma_{L2}^2)^{1/2} \qquad (4.15)$$

The symbols have the same meaning as in Eq. 4.5, except that the subscript numerals indicate the two ears. If both noise components originate in the periphery and are sufficiently small, then, by analogy to Eq. 4.9, we can write approximately

$$\sigma_{L1,2} = a_{1,2} F'_{1,2}(I_{1,2}) \sigma_{I1,2} \tag{4.16}$$

By introducing the approximate expressions for σ_{L1} and σ_{L2} from Eq. 4.16 into Eq. 4.15 and absorbing the constants $a_{1,2}$ into the respective functions, we obtain

$$\sigma_L = [F'^2_1(I_1)\sigma^2_{I1} + F'^2_2(I_2)\sigma^2_{I2}]^{1/2} \tag{4.17}$$

On the assumption that the just detectable loudness difference between the two ears is ΔL and the detection index is $d' = \Delta L / \sigma_{L'}$, we obtain

$$\Delta L - d'[F'^2_1(I_1)\sigma^2_{I1} + F'^2_2(I_2)\sigma^2_{I2}]^{1/2} \tag{4.18}$$

Replacing ΔL by its expression of Eq. 4.7, including the expression $F'_1(I_1)$ in the bracket, and dividing both sides by I_1, we finally arrive at an equation for ear 1, expressed entirely in stimulus terms

$$\frac{\Delta I_1}{I_1} = d' \left\{ \left(\frac{\sigma_{I1}}{I_1}\right)^2 + \left[\frac{F'_2(I_2)}{F'_1(I_1)}\right]^2 \left(\frac{\sigma_{I2}}{I_1}\right)^2 \right\}^{1/2} \tag{4.19}$$

This is the most general form of the Weber-fraction equation for loudness matching and, more generally, sensation-magnitude matching. With appropriate choices of parameter values, it should account for all the empirical jnd relationships.

At magnitude equality, the standard deviations, σ_{I1} and σ_{I2} should be equal, so that Eq. 4.19 takes the simplified form

$$\frac{\Delta I_1}{I_1} = d' \left(\frac{\sigma_{I1}}{I_1}\right) \left\{ 1 + \left[\frac{F'_2(I_2)}{F'_1(I_1)}\right]^2 \right\}^{1/2} \tag{4.20}$$

When the jnds are determined by detecting increments in an ongoing stimulus, only one function of I is present, so that $F'_2(I) = 0$. Generalizing, Eq. 4.20 is reduced to

$$\frac{\Delta I}{I} = d' \left(\frac{\sigma_I}{I}\right) \tag{4.21}$$

Because σ_I tends to be directly proportional to I, and the derivatives, F', do not appear in the equation, the Weber fraction becomes independent of the rate of growth of I and also of I itself. When σ_I grows somewhat more slowly, the near miss to Webers law results.

When the jnds are determined by differentiating between the magnitudes of two increments in an otherwise constant signal, or of two tone bursts of equal sound frequency, $F'_2(I) = F'_1(I)$, Eq. 4.20 is simplified to

$$\frac{\Delta I}{I} = d' \left(\frac{\sigma_I}{I}\right) (2)^{1/2} \tag{4.22}$$

The Weber fraction in Eq. 4.22 for magnitude discrimination between two increments or bursts comes out to be greater by the square root of two than the
Weber fraction determined with single increments. The phenomenon was observed
qualitatively by numerous experimenters and was investigated systematically by
Turner et al. (1989). They used both paradigms on the same group of listeners. The
stimuli consisted of 200-ms bursts of 500, 1,000, or 6,000-Hz tones turned on and
off with 25-ms ramps for the two-stimulus paradigm and of similarly shaped single
increments in 1.6-s pedestals for the single-stimulus paradigm. When their data are
averaged over all the conditions they used, an average ratio of 1.59 results between
the Weber fractions obtained by the two paradigms. The ratio does not appear to
depend systematically on either the sound frequency or the sensation level. It is only
13% greater than theoretically predicted. The difference may be due to temporal
separation of the two-tone bursts, which would introduce a memory factor.

A particularly simple expression for the Weber fraction associated with magnitude matching is obtained when the functions, $F(I)$, are power functions, as they are
likely to be. Then, for the loudness example:

$$\Delta L_{1,2} = \left(\frac{\theta_{1,2} k_{1,2} I_{1,2}^{\theta_{1,2}}}{I_{1,2}} \right) \Delta I_{1,2} \tag{4.23}$$

The latter equation can be simplified to

$$\Delta L_{1,2} = \left(\frac{\theta_{1,2} L_{1,2}}{I_{1,2}} \right) \Delta I_{1,2} \tag{4.24}$$

At loudness equality, $L_1 = L_2 = L$, so that, with Eq. 4.18,

$$\left(\frac{\Delta L}{L} \right) = d' \left[\theta_1^2 \left(\frac{\sigma_{I1}}{I_1} \right)^2 + \theta_2^2 \left(\frac{\sigma_{I2}}{I_2} \right)^2 \right]^{1/2} \tag{4.25}$$

Equation (4.24) can be rewritten for ear one in the form

$$\left(\frac{\Delta L_1}{L_1} \right) = \frac{\theta_1 \Delta I_1}{I_1} \tag{4.26}$$

By replacing $\Delta L/L$ in Eq. 4.25 by its right-hand expression of Eq. 4.26 and by
dividing its both side by θ_1, we arrive at the equation:

$$\left(\frac{\Delta I_1}{I_1} \right) = d' \left[\left(\frac{\sigma_{I1}}{I_1} \right)^2 + \left(\frac{\theta_2^2}{\theta_1^2} \right) \left(\frac{\sigma_{I2}}{I_2} \right)^2 \right]^{1/2} \tag{4.27}$$

This form of the equation for differential sensitivity should be valid not only for
hearing but also for other sense modalities.

Because of the independence of Weber's fraction of stimulus intensity, or nearly
so, and in agreement with the experimental results of Cefaratti and Zwislocki

(1994), we must have approximately $(\sigma_{I1}/I_1) = \sigma_{I2}/I_2$. If this is so, Eq. 4.27 can be simplified for ear 1 in the auditory example to the elegant form,

$$\left(\frac{\Delta I_1}{I_1}\right) = d'\left(\frac{\sigma_{I1}}{I_1}\right)\left[1+\left(\frac{\theta_2}{\theta_1}\right)^2\right]^{1/2} \qquad (4.28)$$

For ear 2, we would then have:

$$\left(\frac{\Delta I_2}{I_2}\right) = d'\left(\frac{\sigma_{I2}}{I_2}\right)\left[1+\left(\frac{\theta_1}{\theta_2}\right)^2\right]^{1/2} \qquad (4.29)$$

Because $(\sigma_{I1}/I_1) = \sigma_{I2}/I_2$, the ratio between the two Weber fractions comes out to be

$$\frac{(\Delta I_1/I_1)}{(\Delta I_2/I_2)} = \frac{\theta_2}{\theta_1} \qquad (4.30)$$

The symbols θ_1 and θ_2 denote the logarithmic slopes of the power functions. Accordingly, in words, the ratio should be inversely equal to the logarithmic slope of the matched functions

The validity of Eqs. (4.28)–(4.30) can be tested on the experimental results of Cefaratti and Zwislocki (1994). For this purpose, Fig. 4.27 shows the loudness-

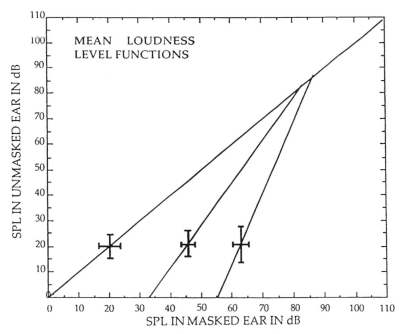

Fig. 4.27 Mean interaural loudness-level curves obtained on three observers at 1 kHz in the absence, respectively, presence of monaural, narrow-band masking noise at two levels. Crossing horizontal and vertical line segments indicate the mean standard deviations of tone level settings in the masked and unmasked ear, respectively. Reproduced from Zwislocki (1996a), by permission of Adam Mickiewicz University Press, Posnan, Poland

matching functions averaged over the individual responses of the three listeners involved in their experiments. Corresponding individual matching functions can be found in Fig. 4.23. In Fig. 4.27, the vertical bars indicate averaged double standard deviations of loudness settings in the unmasked ear. The horizontal bars do the same for the masked ear. Note that the standard deviations in the unmasked ear grow with the slope of the loudness-level function in the contralateral ear, in agreement with Eqs. (4.28) and (4.29), even though the slope in the unmasked ear remains constant. This is perhaps the most striking paradox-like phenomenon predicted by the theory developed above. The standard deviations of loudness matches in the masked ear decrease somewhat with the slope of the loudness-level function in this ear, again in agreement with the theory. Note that the standard deviations referring to the left-most curve in the graph, which has a slope approaching unity, are approximately equal.

For a more explicit validation of the law of differential sensitivity embodied in Eqs. (4.28) and (4.29), the standard deviations of Fig. 4.27, expressed as coefficients of variation, are plotted in Fig. 4.28 by the filled and unfilled circles relative to the slopes of the loudness-matching functions. The corresponding theoretical predictions, normalized to the coefficient of variation at $\theta_{2,1} = 0$, are plotted as the solid

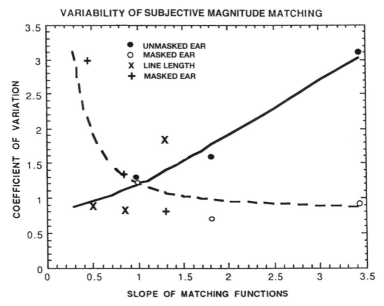

Fig. 4.28 Coefficients of variation (Weber fractions) of stimulus-magnitude adjustments for subjective-magnitude equality versus the slope of the matching curves. Circles refer to intramodal sound-intensity adjustments, crosses, to intermodal sound-intensity and line-length adjustments. Unfilled circles and vertical crosses hold for the masked ear, filled circles and slanted crosses for the unmasked ear and the line length, respectively. The curves are theoretical, the intermittent curve referring to the masked ear, the solid curve, to the unmasked ear and the line length, respectively. Data from Cefaratti and Zwislocki (1994), curves from Zwislocki (1996a). Reproduced by permission of Adam Mickiewicz University Press, Posnan, Poland

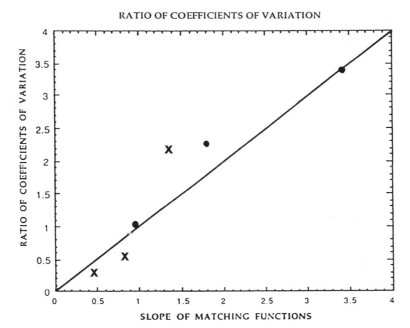

Fig. 4.29 Ratios of the coefficients of variation (Weber fractions) for stimulus-magnitude adjustments in the masked and unmasked ears (filled circles); and in the masked ear and line length (slanted crosses). The diagonal line is theoretical. It is based on the assumption that the noise coefficients of variation, σ/I, associated with all the variables involved (tone in masked and unmasked ears and line length) are equal. Based on the data of Cefaratti and Zwislocki (1994). Reproduced by permission of Adam Mickiewicz University Press, Posnan, Poland

and intermittent lines. To broaden the scope of validation to intermodal magnitude matching, the averaged coefficients of variation resulting from the experiment on matching line length to loudness and reported in Fig. 4.26 are also included in terms of slanted and upright crosses. Both sets of the data points, intramodal and intermodal, follow grossly the theoretical curves. Not surprisingly, the intermodal points show greater deviations than the intramodal ones. Referring to the intramodal data points, note the double circle near the slope of unity of the matching functions. It indicates a value of the coefficient of variation greater than its value at $\theta_{2,1} = 0$ by approximately a square root of two. Remember that the coefficients of variation are equivalent here to Weber fractions and that the Weber fraction obtained in experiments based on comparison of magnitudes of two stimulus bursts, or two increments in an ongoing stimulus, tends to be greater than the fraction obtained in experiments based on detection of single increments by the square root of two.

A further validation of the relationships described in Eqs. (4.28) and (4.29), can be obtained by taking their quotient embodied in Eq. 4.30 on the assumption that $(\sigma_{I1}/I_1) = \sigma_{I2}/I_2$. The two fractions have the meaning of coefficients of variation equal to Weber fractions that result from the detection of increments in a continuous stimulus and the detection index, $d' = 1$. The equation is displayed graphically in

Fig. 4.29 by the solid line. The filled circles show the experimental results derived from intramodal loudness matching, the crosses do the same for the intermodal matching of line-length and loudness. There is fair agreement between the theoretical predictions and the experimental data, except for two idiosyncratic points at medium slopes of the magnitude-matching functions. They correspond to the most deviant points in Fig. 4.28. Further experiments, performed on a larger number of observers would be required to decide if the deviations are purely statistical or have phenomenological significance.

Summarizing, I feel justified in stating that Eq. 4.19 and its simplified form for power functions, Eq. 4.29, adequately describe the Weber fractions resulting from all three types of experiments – those based on single intensity increments, those based on an intensity difference between two stimulus bursts or two increments in an ongoing stimulus, and those based on magnitude matching. Therefore, the equations embody what I have called the *general law of differential sensitivity*.

References

Cefaratti, L.K. and Zwislocki, J.J. (1994). Relationships between the variability of magnitude matching and the slope of magnitude level functions. *Journal of the Acoustical Society of America* 96(1); 126–133.

Doerfler, L.G. (1948). Differential sensitivity to intensity in the perceptually deafened ear. *Ph.D. Dissertation*, Northwestern University, Evanston, IL.

Engen, T. (1972). Psychophysics I: Discrimination and detection. In: J.W. Kling and L.A. Riggs (Eds.), *Woodworth & Schlosberg's Experimental Psychology, 3rd Edition: Vol. 1, Sensation and Perception*. Holt, Rinehart & Winston, New York, pp. 11–46.

Gescheider, G.A., Bolanowski, S.J., Verrillo, R., Arpajian, D.J., and Ryan, T.F. (1990). Vibrotactile intensity discrimination measured by three methods. *Journal of the Acoustical Society of America* 87(1); 330–338.

Gescheider, G.A., Bolanowski, S.J., Zwislocki, J.J., Hall, K.L., and Mascia, C. (1994). The effects of masking on the growth of vibrotactile sensation magnitude and on the amplitude difference limen: A test of the equal sensation magnitude-equal difference limen hypothesis. *Journal of the Acoustical Society of America* 96(3); 1479–1488.

Hecht, S. (1934). Vision II. The nature of the photoreceptor process. In: C. Murchison (Ed.), *Handbook of General Experimental Psychology*. Clark University Press, Worchester.

Hellman, R., Scharf, B., Teghtsoonian, M., and Teghtsoonian, R. (1987). On the relation between the growth of loudness and the discrimination of intensity for pure tones. *Journal of the Acoustical Society of America* 82(2); 448–453.

Jesteadt, W. (1980). An adaptive procedure for subjective judgments. *Perception & Psychophysics* 28(1); 85–88.

Jesteadt, W., Wier, C.G., and Green, D.M. (1977). Intensity discrimination as a function of frequency and sensation level. *Journal of the Acoustical Society of America* 61; 169–177.

Johnson, J.H., Turner, C.W., Zwislocki, J.J., and Margolis, R.H. (1993). Just noticeable differences for intensity and their relation to loudness. *Journal of the Acoustical Society of America* 93(2); 983–991.

Jordan, H.N. (1962). An investigation of the validity and reliability of the Lüscher-Zwislocki test of loudness recruitment. *Ph.D. Dissertation*, Syracuse University, Syracuse, NY.

König, A. and Brodhun, E. (1889). Experimentelle Untersuchungen ueber die psychophysische Fundamentalformel in Bezug auf den Gesichtssinn. *Sitzungsberichte Preussische Akademie Wissenschhaften* 27; 641–644.

Levitt, H. (1971). Transformed up-down methods in psychoacoustics. *Journal of the Acoustical Society of America* 49; 467–477.

Licklider, J.C.R. (1951). Basic correlates of the auditory stimulus. In: S.S. Stevens (Ed.), *Handbook of Experimental Psychology*. Wiley, New York, pp. 985–1039.

Lüscher, E. and Zwislocki, J.J. (1948). A simple method for indirect monaural determination of the recruitment phenomenon. *Acta Otolaryngologica Supplement* 78; 156–168.

Lüscher E. and Zwislocki, J.J. (1951). Comparison of the various methods employed in the determination of the recruitment phenomenon. *Journal of Laryngology and Otology* 65; 187–195.

McGill, W.J. and Goldberg, J.P. (1968). Pure-tone intensity discrimination and energy detection. *Journal of the Acoustical Society of America* 44; 576–581.

Miller, G.A. (1947). Sensitivity to changes in the intensity of white noise and its relation to masking and loudness. *Journal of the Acoustical Society of America* 19; 609–619.

Ozimek, E. and Zwislocki, J. J. (1996). Relationships of intensity discrimination to sensation and loudness levels: Dependence on sound frequency. *Journal of the Acoustical Society of America* 100; 3304–3320.

Penner, M.J., Leshowitz, B., Cudahy, E., and Richards G. (1974). Intensity discrimination for pulsed sinusoids of various frequencies. *Perception and Psychophysics* 15; 568–570.

Riesz, R.R. (1928). Differential sensitivity of the ear for pure tones. *Physics Review* 31; 867–875.

Riggs, L.A. (1972). Vision. In: J.W. Kling and L.A. Riggs (Eds.), *Woodworth & Schlosberg's Experimental Psychology, 3rd Edition: Vol. 1, Sensation and Perception*. Holt, Rinehart & Winston, New York, pp. 273–314.

Schacknow, P.N. and Raab, D.H. (1973). Intensity discrimination of tone bursts and the form of the Weber function. *Perception and Psychophysics* 14; 449–450.

Stellmack, M.A., Viemeister, N.F., and Byrne, A.J. (2004). Monaural and interaural intensity discrimination: Level effects and the "binaural advantage." *Journal of the Acoustical Society of America* 116(2); 1149–1159.

Stillman, J.A., Zwislocki, J.J., Zhang, M., and Cefaratti, L.K. (1993). Intensity just-noticeable differences at equal-loudness levels in normal and pathological ears. *Journal of the Acoustical Society of America* 93(1); 425–434.

Turner, C.W., Zwislocki, J.J., and Filion, P.R. (1989). Intensity discrimination determined with two paradigms in normal and hearing-impaired subjects. *Journal of the Acoustical Society of America* 86(1); 109–115.

Zwislocki, J.J. (1996a). A theory for intensity jnds and the variability of subjective magnitude matches. In: U. Jorasz, (Ed.), *Auditory Perception: Principles and Applications*. World Scientific Publishing Co., Singapore.

Zwislocki, J.J. (1996b). Cochlear mechanisms underlying loudness recruitment and the half-octave shift in cochlear noise damage. *Central & East European Journal of Otorhinolaryngology, Head and Neck Surgery* 1, 6–18, 1996.

Zwislocki, J.J. and Jordan, H.N. (1986). On the relations of intensity jnds to loudness and neural noise. *Journal of the Acoustical Society of America* 79(3); 772–780.

Zwislocki, J.J. and Relkin, E.M. (2001). On a psychological transformed-rule up and down method converging on a 75% level of correct responses. *Proceedings of the National Academy of Science* 98(8); 4811–4814.

Zwislocki, J.J., Maire, F., Feldman, A.S., and Rubin, H. (1958). On the effect of practice and motivation on the threshold of audibility. *Journal of the Acoustical Society of America* 30; 254–262.

Subject Index

Author Index